The Future of Pricing

The Future of Pricing

How Airline Ticket Pricing Has Inspired a Revolution

E. Andrew Boyd

palgrave
macmillan

THE FUTURE OF PRICING
Copyright © E. Andrew Boyd, 2007.

First published in 2007 by
PALGRAVE MACMILLAN™
175 Fifth Avenue, New York, N.Y. 10010 and
Houndmills, Basingstoke, Hampshire, England RG21 6XS.
Companies and representatives throughout the world.

PALGRAVE MACMILLAN is the global academic imprint of the Palgrave Macmillan division of St. Martin's Press, LLC and of Palgrave Macmillan Ltd. Macmillan® is a registered trademark in the United States, United Kingdom and other countries. Palgrave is a registered trademark in the European Union and other countries.

ISBN-13: 978-0-230-60019-5
ISBN-10: 0-230-60019-0

Library of Congress Cataloging-in-Publication Data
Boyd, E. Andrew.
The future of pricing : how airline ticket pricing has inspired a revolution / E. Andrew Boyd.
    p. cm.
Includes bibliographical references and index.
ISBN 0-230-60019-0 (alk. paper)
1. Airlines—Rates. 2. Airlines—Prices. 3. Transportation—Tickets. 4. Pricing. I. Title.
HE9783.5.B69 2007
387.712—dc22   2007003837

A catalogue record of the book is available from the British Library.

Design by Scribe, Inc.

First edition: Novemeber 2007

10 9 8 7 6 5 4 3 2 1

Printed in the United States of America.

For my sister, mother, and father.

Contents

Acknowledgments   ix

Preface   xi

1   A Revolution in the Making   1

2   Method or Madness?   7

3   The Computer Did It   23

4   How It All Works   43

5   When Passengers Collide   57

6   Hold Me, Darlin'   67

7   Upon Arrival: Hotels, Rental Cars, Cruise Ships, and More   85

8   The Just Price   103

9   The Scientists   117

10 The State of Pricing   135

11 Pricing's Many Faces   147

12 The Coming Revolution   165

Appendix   177

Reference List   179

Index   187

Acknowledgments

The agents I spoke with while preparing *The Future of Pricing* wanted me to write either a professional memoir or a "guru" book—a book in which I presented myself as a leading authority on pricing. A memoir was inappropriate because the goal wasn't to tell a story about my professional life. A guru book was out as well. I'm not a guru. I'm an observer and participant who was fortunate enough to find himself in a fascinating place at a fascinating time. While I like to think I've made valuable contributions to the field of pricing—both mathematically and managerially—scientific pricing is the result of the contributions of many different individuals. (I *can* thank the agents for convincing me that it was okay to use "I" in a sentence and to share personal experiences.)

I would like to thank every pricing and revenue management professional I've worked with, conversed with, or read the writings of. Each has contributed to this book in his or her own way. While it would be impossible to list everyone by name, there are a few individuals I would like to recognize explicitly: Chris Anderson, Peter Bell, Peter Belobaba, Bill Cooper, Bob Cross, Renwick Curry, Guillermo Gallego, Tito Homem-de-Mello, Itir Karaesmen, Sherri Kimes, Anton Kleywegt, Warren Lieberman, Jeff McGill, Georgia Perakis, Laura Preslan, Bob Phillips, Richard Ratliff, Rob Shumsky, Barry Smith, Larry Weatherford, and Ciyou Zhu. Special thanks to Karl Isler, Kalyan Talluri, and Garrett van Ryzin, all of whom had a tremendous impact on shaping my thoughts.

Many people contributed to this book by providing information or sitting for an interview. Thanks to Steve Bondi, Bill Brunger, Tom Cook, Neil Corbett, Dave Cripe, Lee Davis, J. D. Dick, Andy Doyle, Bill Dudziak, John Evler, Stacy Foree, Daryn Herrington, Bill Hinke, Phi Hoang, Pete Horner, Jerry Jung, Jeff Kabin, Leon Kinloch, Steve Klein, Vern Lennon, Barry List, Greg Lough, Al Ludwig, Irv Lustig, Steve Macadam, Pat McCord, Chuck McElrea, Denise Minier, Steve Neubert, Graham Parker, Steve Pinchuk, Todd Plage, John Quillinan, Beju Rao, John Riddell, Mike Rothkopf, Mark Shafer, Martin Simoncic, Manny Sousa, Benson Yuen, and Jim Ziminski. Special thanks to

Bob Salter, who provided extensive background material and support throughout the writing process and who, together with Neil Biehn, read early drafts of the manuscript.

Closer to home, I'd like to thank the many scientists at PROS past and present, without whom the Science and Research Group could never have evolved to what it is. Thanks, too, to the entire PROS organization. From top to bottom, it's a team I enjoy working with and am proud to be part of. Growing a company from a handful of people to one of the three largest software firms in the city of Houston is challenging, but for all the ups and downs, we've managed to keep our humor and maintain a culture in which we care about what we do. As CEO Bert Winemiller puts it, "do the right thing." I repeat this often. Thanks, Bert. Thanks, too, to Charlie Murphy for serving as the yin to Bert's yang. And special thanks to Ron Woestemeyer and Mariette Melchior. Mariette, you lured me away from an academic career and I'm glad you did. Life at PROS has been way too exciting to miss.

Acquisitions editor Aaron Javsicas was a pleasure to work with, and I'd like to thank both Aaron and Palgrave Macmillan for providing me with the freedom to write a book that falls somewhat outside the domain of a traditional academic press. Thanks as well to Kate Ankofski, Chris Chappell, Dale Rohrbaugh, Erica Warren, and the entire editorial staff at Palgrave Macmillan for their help and support.

Finally, I'd like to thank my son, Alex, and daughter, Katy, who are my constant source of joy. But most of all, I want to thank my wife, Sarah, who has always been there for me. S., life just wouldn't be the same without you.

Preface

Every occupation has its unwritten laws of professional decorum. Having spent eight years as a university student and another ten as a professor, I'm familiar with the requirements of general academic writing: third person passive voice, using an idiom familiar to a specialized group, conveying a serious tone. And for my chosen profession, I add the use of mathematics. If an academic researcher wants to advance in their career, there are rules to be followed.

Having now spent more time in business than as a professor, it's fair to say that my writing style no longer conforms to academic protocols. Out of necessity and, admittedly, desire, I've become comfortable with simpler, more straightforward writing, since I've discovered that it's often far more effective at getting your point across. What you give up in precision, you easily make up for in clarity and comprehension. This book strays from academic conventions, but it does so purposefully.

The Future of Pricing is intended to teach, but not through formulas or, in most instances, explicit arguments leading to explicit conclusions. Instead, I've used interviews, examples, history, and personal experiences to make my case. There's a carefully thought out thesis, but I intentionally refrained from using the compartmentalized structure found in textbooks and how-to books. I sincerely hope you walk away understanding what incredible changes are taking place in pricing, and for those who are interested, a sense of the marvelous mathematics behind airline ticket pricing. I also hope that you come to better understand the interplay between science, technology, and people in a business setting, and the challenges faced by organizations seeking to integrate science into their pricing. But I also hope you find the book enjoyable, too.

CHAPTER 1

A Revolution in the Making

What's the right price for a bar of soap? Nothing fancy, just the soap we reach for in the shower every morning, or bathe our children with before we put them to bed.

As consumers, we want a price that's as low as possible. If we found a bar of our favorite soap for sale at the local superstore for thirty cents, we'd be elated. But we'd be happier if the price was twenty cents, and even happier if it was only ten.

It's also true that soap retailers and manufacturers want a price that's as high as possible. But unlike consumers, they're expected to set the price. This is a more difficult task. No longer is the decision "yes, I'll take it" or "no, I'll pass," but rather, "how much should I ask someone to pay?"

While most of us give little thought to how prices are actually set, any company that provides goods or services must constantly worry about price. If it can't set prices high enough to cover costs, it will go out of business. If it can set prices high enough to generate a reasonable margin, bonuses will flow and shareholders will be rewarded with dividends and an increased stock price.

Given the importance of price, we might expect that companies have always devoted immense time and effort to pricing. While it's true that companies spend considerable time worrying about price, the reality is that most determine their prices using less-than-ingenious means. Sometimes prices are set by adding a small margin to costs, paying little or no attention to what the market is willing to bear. In other cases, longtime industry veterans use rules of thumb developed through experience. All too often, companies simply look at what competitors are doing and do the same thing. These responses are understandable given the difficulty of the pricing problem. It's easy to imagine an ideal price residing in some Platonic stratum. It's another issue to actually find it. A retail pricer once confided to me that he used a "ceiling" strategy when forecasting future demand for setting prices. "I lean back in my chair and look at the ceiling," he said, "and I generate a forecast."

Retail pricing is tough enough, but in cases in which sales agents are involved, the problem is often more difficult. Faced with commission structures based on volume, and responding to a human need to avoid unwanted conflict, sales agents endeavor to get the lowest possible price for their customers, or at least a fair price, however they arrive at their notion of what is fair. When price approval is required, they negotiate with pricing managers who are expected to toe the line on price but who are often just as worried about losing a sale as the agent is. Sales agents view pricing managers as condescending and out of touch with reality, while pricing managers view agents as too willing to take customers golfing and too unwilling to stand up and ask for a good price. Although managers may have MBAs, many sales agents have less—or a less specialized—education, a fact that can further strain the relationship.

Piles of paper are exchanged up and down the pricing hierarchy as special deals receive approval or are sent back for revision. At one company, the process was so complicated that they sent the wrong price list to their biggest customer. As a result, the company lost the business to a competitor. Realizing what had happened, company executives apologetically approached the customer and were able to win back the business, but only after additional price concessions that cost millions of dollars. Even in companies that have adopted electronic means for managing the pricing process, spreadsheets rule, and a coordinated, logical flow of documents and activities rarely exists.

The largely unscientific means by which most companies price is especially remarkable given the available technology for gathering and storing information about what's been sold. Scanners read price tags and automatically update inventory information. Some even place orders for new inventory automatically. Web sites process purchases with the click of a mouse, electronically triggering a sequence of events that leads to a delivery on your doorstep in a matter of days. Technology is available to make better pricing decisions, yet most industries are only beginning to take advantage of it. Recognizing an opportunity and capitalizing on that opportunity, however, are entirely different problems. And as companies seek to improve their pricing, they look for successful examples they can emulate.

By far one of the most successful examples is that of operational airline pricing. For almost a century, airlines have honed their operational pricing skills. The airline industry was remarkably unsophisticated in the early days of flying, and airline history is rife with freewheeling pilots and cigar smoking executives who played power politics during many long years of regulation. Yet airlines were among the first businesses to set aside pencil and paper and embrace computers, entering the world of electronic sales a decade before the invention of the integrated circuit. Today we would consider their early efforts comical, but they were on the cutting edge of technology as they experimented with record-keeping gumball machines and "Magnetronic Reservisors," the latter of which are on display at the Smithsonian Institution. These mechanical devices set the stage for the development of computer reservations systems and played a central role in the ferocious price battles that occurred in the wake of deregulation. Fighting with the first electronic distribution networks, airlines experienced firsthand the power and pitfalls of e-commerce years before the arrival of the Internet.

Today, carriers including American, United, Lufthansa German Airlines, and Swiss employ staffs of PhDs trained in mathematics to help solve their pricing problems. Others rely on PhDs working for software vendors and consultants. Computers sort though terabytes of data containing information about passenger booking behavior on a nightly basis, allowing prices to be adjusted with every booking or cancellation. By almost any measure, airlines practice the most sophisticated pricing ever conceived.

The flying public might not agree. At times, airline pricing just seems nutty, and it can certainly be frustrating. We all have our favorite stories of the great deal we uncovered or the time we were forced to pay an outrageous sum for the privilege of getting from point A to point B. Our stories are laced with middle seats, first class upgrades, and frequent flyer miles. Flying adventures have supplanted weather as the ubiquitous icebreaking topic at the conference room table as businesspeople gather from around the world to start the day's work.

I began my journey into the airline industry on January 1, 1997. Holding a PhD from MIT and working as a tenured faculty member in the nationally ranked industrial engineering department at Texas A&M University, restlessness got the better of me. I enjoyed studying the facial structure of million-dimensional geometric figures, and I still do, but my work was becoming increasingly narrow and distant from the problems it was intended to solve. When a former student working in a small and growing company named PROS brought to my attention an opportunity to build a science group, I jumped. The National Science Foundation had long been preaching the value of getting professors into the workforce to gain practical experience, and I decided it was time to heed the message.

What I discovered was beyond my imagination. Within the first week of my arrival, I was on a plane to visit Northwest Airlines at their headquarters in Minneapolis, a place most people attempt to flee during the winter months. After a dark, cold ride during which the cab driver explained the joys of ice fishing in an accent made famous in the movie *Fargo*, I found myself ushered into a warm, brightly lit room. My job was to help explain a forecasting system I knew nothing about to an industry I had no experience with. Fortunately for all involved, I was accompanied by an experienced sales and technical team that was quite knowledgeable about the system.

Faced with a vice president, director, operational pricing analysts, and Northwest's own set of PhDs, the discussion centered on the relative merits of different forecasting algorithms embedded in a language of AUs (authorization levels), DCPs (data collection points), and other strange terms. I would soon learn that dialects of this same language are spoken by operational pricing groups throughout the travel industry. Deciphering the jargon would come with time, but from that very first meeting, I was amazed by the level of mathematical sophistication during the discussion. Had I just been speaking scientist to scientist, the discussion would have come as no surprise. But executives don't typically care about elaborate science. They care about staying in business. Yet here were executives who *did* care about science because it was necessary for their survival.

My perception was only furthered as I traveled the globe: Frankfurt, Hong Kong, Montreal, Singapore, Zurich; I quickly discovered the realities of what it means to work with a global industry. As chief scientist, my primary duty was to make sure the pricing science embedded in PROS's software brought the highest possible value to our clients, but it was also my job to support the sales organization during competitive high-end sales visits to major airlines. As the second millennium came to a close, the demand for sophisticated pricing science was reaching a crescendo. Thrown into a sales presentation, a person off the street would not have had a clue what the discussion was about. Bayesian hierarchical forecasting and dual multipliers from constrained optimization problems—I could only marvel that the battle to sell pricing software to one of the world's great industries was being waged on a field of lambdas and sigmas. These were signs of an incredibly mature industry, one that had grappled with pricing problems for years.

While I reveled in the opportunity to extend the mathematical boundaries of pricing science even further, I became increasingly interested in how things had ever evolved to such a point in the first place. Even more, I wondered what the history of airline ticket pricing foretold about the future of pricing in general. Would we one day find our bar of soap changing price on a minute by minute basis, the result of detailed mathematical algorithms conceived by some of the world's best educated minds? Would sales agents use computers not just to look up prices set by a committee, but also to calculate prices based on historical trends, inventory levels, market demand, and the characteristics of the customer being sold to?

These and similar questions have yet to be answered, but they are being asked with a new sense of urgency by all businesses. Business interest springs from three primary sources. First is a renewed realization of the leverage pricing has on bottom line profits, since small percentage changes in price can lead to large percentage changes in profitability. A company selling a product for $101 that costs $100 to manufacture can double its profitability if it can sell the product for $102 without drastically reducing demand. A growing number of successful pricing initiatives are appearing in industries in which scientific pricing hasn't been the norm. In many cases, these efforts were inspired by experiences with airline pricing or were actually sponsored by former airline executives who have moved on to new positions.

Second is the availability of data—all kinds of data. Science thrives on data. In the early 1990s, corporations began making large investments in enterprise resource planning software to manage the basic data involved in running a business. Customer relationship management systems went one step further, gathering detailed data about individual transactions and customers. Your family enjoyed their stay in the ocean view room last year? Let me offer something similar this year. The Internet unlocked the data floodgates, providing a means to experiment and proactively gather information on what sells.

Third, many business leaders realize they are simply not doing a good job of pricing. At one company, management was concerned that their workforce of telephone sales agents were not holding the line on prices. Customers used all sorts of ploys to negotiate the lowest possible price, and, unwilling to risk not making a sale, the sales agents would routinely acquiesce. It didn't matter that the item under consideration was almost out of stock and would not be replenished any time in the near future. Nor did it matter that the market price had been increasing for days. Without this information and without projections on what the market would bear, the sales agents would rely on whatever habits they had developed over the years. Provided with the right information and the right incentives, it was not difficult to raise overall profitability by double-digit percentages.

The spectrum of industries changing the way they look at pricing is remarkable: energy, chemicals, cargo, financial, pharmaceuticals, retail, consumer packaged goods, high-tech manufacturers, advertising, storage, even hospitals; the list goes on and on. With many decades of gathering and storing data, and with most of the back office applications automated, these companies are hungry to do something with their investment. Some are ready to push the envelope, taking the cumulative knowledge of the past and making the organizational and computer system changes necessary to bring about advanced scientific pricing today. However, many companies are in the same position as the airline industry thirty years ago. They can see the value of scientific pricing, but they are not quite sure about it. How will it affect my business? Is my company ready? How hard will it be?

Pricing has been with us throughout history, undergoing many changes along the way. Yet for the most part, it still relies on instinct and intuition because of a deeply held belief that instinct and intuition are correct. In an era when computers are capable of capturing and analyzing individual transactions, our beliefs are now being put to the test. And as they are, we will increasingly find that we are just not able to juggle the data with the adroitness of our mechanical counterparts. From understanding how people behave, to setting and managing prices in real time at the micromarket level, science will play an ever-increasing part in determining what we pay. Humans will never be extricated from their role in pricing, but that role will change as science-based computer systems become extensions of trained analysts. It's still too early to know exactly what scientific pricing will look like for the many different pricing problems faced by industry. Given the subculture of pricing found in most organizations, change is not likely to occur overnight. But scientific pricing is already occurring all around us, just as it was for the airlines in the mid-1970s.

CHAPTER 2

Method or Madness?

It's not a hobby, really, nor a preoccupation. It's an all-out obsession.

Benson Yuen loves airplanes. The love affair began as a child in Hong Kong, where from his first trip to the airport, he was hooked. But unlike most children, Yuen didn't see himself in the cockpit. Instead, it was the way the whole system came together that intrigued him. Aircraft, strangely shaped trucks, flight crews, ticket agents, baggage handlers, caterers—all bringing together people who might otherwise never see each other. To Yuen, it was so modern, mechanical, and yet human.

After moving to the United States, one of Yuen's favorite childhood pastimes was visiting Kennedy Airport to collect flight schedules, then study and memorize them. It's no surprise that in a paper for an upper level college English course, he examined how Pan Am was beginning to fragment following its National Airlines acquisition. The teacher, recognizing Yuen's passion, passed along a job advertisement for a new airline forming in Florida. Yuen landed an interview with four senior executives of the company who also happened to be the only four employees at the time.

Although it had no airplanes and little capital, Florida Express did have experienced leadership from the likes of Pan Am, Frontier, and Midway. A few weeks later, although he was highly overqualified, Yuen held the title of office manager, which in a company of five people translated into "do anything the other four guys need to have done." But within a few years, the company had grown to more than two-dozen aircraft, and Yuen found himself at the center of market planning, scheduling, and most importantly for his future life, inventory control.

Inventory control is the heart of pricing in the airline industry. Airlines create different products out of the same basic resource—a seat on an aircraft—by attaching different restrictions. A low priced product may require a Saturday night stay and carry change fees, while a high priced product may carry no restrictions and be fully refundable. These products, almost universally represented by a single alphabetic character such as Y or Q, are assigned prices by an airline's pricing department.

If airlines simply put their products on a shelf and let passengers buy what they want until the seats run out, inventory control would consist of counting seats as they're sold. But over time, the airlines realized there was no need to keep the shelf fully stocked. If a lot of people are willing to pay for the high-priced product, why not simply stop selling the low-priced product? This is what airline inventory control is all about—deciding how much of each product to make available. The price changes we see when buying an airline ticket are far more often due to what's on the shelf at any given moment than actual price changes. We see airline inventory control as *operational pricing* or simply *pricing*, however, because we experience it in terms of price changes.

In 1987, five years after joining Florida Express, Yuen interviewed with Ronald Woestemeyer and Robert Salter at Propsys, a small company that did specialized work for the airlines. Eastern Airlines had waged a vicious price war with Florida Express. Florida Express was being sold, and one of Yuen's mentors encouraged him to speak with the two entrepreneurs.

Two decades earlier, Woestemeyer and Salter earned their wings working for Frank Lorenzo, chief executive at Texas International Airlines. Lorenzo later purchased Continental and a host of other airlines before renaming the resultant company Continental. Salter encouraged Woestemeyer, his fraternity brother at the University of Houston, to leave his position at Phillip Morris and join Texas International. Both held a variety of positions in sales and marketing throughout their careers at the airline. In spite of the many challenges they faced, Woestemeyer and Salter look back fondly at that period of their lives. It was a chance for two young men to learn about the industry and—thanks to the many empty seats on aircraft at that time—see the world.

Woestemeyer left the airline in 1982 to focus his attention on a newly evolving practice known as *yield management*, later known as *revenue management*. Revenue management seeks to increase the average revenue passengers generated on each plane by applying forecasting and optimization science to manage what products should be available for sale on any given day and in what quantity. Woestemeyer returned the invitation from many years earlier by asking Salter to join him at Propsys, an offer Salter accepted.

The opportunity to join Woestemeyer and Salter proved perfect for Yuen, who again became one of a handful of entrepreneurs looking to make their mark on the airline industry. But this time was different. The team wasn't running an airline, they were providing a service to many different airlines. And while airlines dealt with inventory control on a daily basis, the idea of using scientific software to do the job wasn't an established practice. Woestemeyer, Salter, and Yuen saw a market to be made, and this required visiting potential customers.

* * *

Travel takes on a new meaning when speaking of the airline industry. Airline headquarters are situated around the globe rather than concentrated in a few major cities. Apart from the fact that this just makes sense for an industry that ferries people from one place to another, running an airline is a source of national pride for many countries. The International Air Transport Association, an industry trade organization whose member airlines carry 94 percent of all scheduled air traffic flown internationally, boasts some 265 members, only 17 of which make their home in the United States or Canada. Names like Singapore, Emirates, and Philippine Airlines dominate the list.

Further, people in the airline industry treat planes the way most people treat taxicabs. It's not unusual for airline personnel to grab a quick 11 hour flight from Houston to Frankfurt, get off the plane during morning rush hour, work for the day, eat a heavy German dinner, and, after a quick night of sleep, hop on another flight back to Houston. Sometimes the night of sleep is foregone. One of Yuen's travel companions logged over 500 thousand miles in one year—roughly the equivalent of flying around the world every two weeks. With so much time in the air, he once complained to Yuen that he'd lost the ability to sleep in a horizontal position—a problem the new generation of business class seats has presumably corrected. For Yuen's part, over the 18 years he's been preaching revenue management, he's averaged 250 thousand miles each year. This is not quite the peak reached by his colleague, but remarkable nonetheless.

Yuen prefers to book his own flights. He does this in part because of his childhood fascination with the way airlines work, but also because no office travel coordinator could ever possibly know what flights best suit his needs. In this regard, Yuen isn't alone. Every day, millions of us from around the world take to the Internet in search of airline tickets. Like Yuen, we're not satisfied with finding a good flight, we want the best. We spend time searching the Web for the best schedule, the best frequent flyer perks, and above all, the best price.

Price is the common denominator. The flight that leaves at nine o'clock in the morning may give us a few hours of extra sleep, but the seven o'clock flight doesn't look so bad if we can save $150. Even business travelers, wed as they are to frequent flyer perks, aren't immune to price. Corporate policy may restrict travelers to flights at or near the lowest price, and many travelers simply feel the desire to win the game or to impose order on a seemingly chaotic world.

We all have our favorite tale about airline ticket prices. Whether it's about a fare that doubled overnight, or was many times lower (or higher) than the person sitting next to us, we are compelled to share our stories over and over. They seem to defy rationality and point to what must be fundamental flaws in the whole pricing process. If we found a video camera that we liked for $400, and went back the following day to find that the price was $600, we'd be stunned. Yet price changes of this type are common for airline tickets. How do such discrepancies come about? Are airlines foolish, out of control, or is there method to the madness?

Ticket pricing isn't much different than pricing video cameras, but it does have its own set of peculiarities. Video cameras have a long shelf life, but airline seats are perishable. They have value right up to the moment the door on the aircraft is closed. After that, empty seats lose their ability to generate revenue. A traveler can exchange tickets, sometimes incurring a penalty, but can't resell them. Airline seats don't need to be delivered.

The biggest difference is that a store stocking two different models of camera can't magically change one into the other. There are two different models, and each has its own price. Airlines, on the other hand, can easily change one product into another just by making a decision to do so. A coach seat works equally well as a Y- or a Q-class product.

Herein lies the multimillion-dollar question. In 2004, Continental Airlines operated over 2,500 daily departures while incurring an operating loss of $230 million. Had the carrier been able to sell one passenger on each flight a ticket worth $300 more than they paid, it would have turned the $230 million loss into a $44 million profit. To help put this into perspective, the union concessions Continental secured in the first half of 2005 were expected to save $418 million annually—about the same as getting one passenger on each flight to pay an additional $450.

Today's airlines are adept at the intricacies of using science to manage fare product availability and overbooking levels on a constant basis, providing a tremendous financial benefit. Building upon the history of work done by the airlines, a recent award-winning academic treatise included a bibliography of some 591 presentations, papers, and books on the topic of inventory control and pricing. It contains chapters filled with equations on network capacity control, overbooking, dynamic pricing, forecasting, and customer behavior. Leading academic institutions such as the Massachusetts Institute of Technology, Columbia, Stanford, Cornell, and the Georgia Institute of Technology now include classroom material on inventory control and pricing, and the topic has become a dominant theme at academic conferences. A search of presentations at the annual gathering of the Institute for Operations Research and the Management Sciences (INFORMS) produced 320 titles containing the words "price," "pricing," or "revenue management."

Research into pricing has grown at a remarkable pace in recent years as academicians seek to understand the best way to price everything from insurance contracts to groceries. The impetus for this activity has its roots in the early successes experienced by airlines and, somewhat later, other parts of the travel industry. The extent to which the travel industry has led the way can be seen in a conference track entitled "Pricing and Revenue Management for Non-Travel Industries." There's the travel industry, then there's everything else.

With decades of accumulated knowledge, airlines continue to play a central role in developing the science of inventory control and pricing. Many of the early problems encountered by airlines are showing up once again as new industries seek to adopt a more structured, scientific approach to pricing. While airlines are quite sophisticated today, it wasn't so long ago that they were grappling with very basic issues, most of which had nothing to do with science.

* * *

The realization that controlling fare class inventories was really the same thing as pricing occurred slowly. From the earliest days, airlines employed reservations agents to take bookings. A passenger or travel agent would call an airline and a reservations agent would let the caller know what flights were available that suited the traveler's needs. If the traveler decided to make a purchase, the reservations agent would dutifully write down the transaction. Airline reservations agents weren't sales people, they were order-takers. Their performance was measured by the number of phone calls they handled.

Under regulation, this was a perfectly viable strategy. Prior to 1978, the Civil Aeronautics Board regulated the routes, schedules, and prices of every airline. No matter who you were or what carrier you flew with, you paid the same price.

There were some exceptions. Airlines were allowed to offer a fare class with a reduced price on flights with unpopular schedules, such as those that flew overnight—the so-called "red-eye" flights. Airlines were also able to offer student and senior discounts. Although the options were limited, they did provide airlines with a first taste of what could be done with inventory control. Even if the Civil Aeronautics Board granted a carrier the right to offer a student discount fare, the carrier wasn't obliged to do so.

Deregulation brought new focus to every aspect of the way airlines attracted passengers. After years of competing on anything but price, airlines were given free rein to determine what products they wanted to sell, at what price, and in what quantity. Airlines could have continued to offer a single product with discounts for special groups. But airline veterans had long been aware that there are two very different types of passengers: business and leisure. Leisure passengers are price sensitive, willing to adjust their schedules to save a few dollars. Business passengers, on the other hand, are far less price sensitive. Making meetings and minimizing time on the road are as important as price—especially if someone else is paying the bill. If there's a single theme that has shaped airline pricing and fare products throughout the years more than any other, it's the distinction between these two types of passengers.

The problem with a single fare is that it can't account for both types of passengers. If 50 people are willing to pay up to $200 for a seat on a plane, and another 50 are willing to pay only $100, the airline is faced with a dilemma. Setting a price of $200, the airline generates: $50 \times \$200 = \$10,000$. At a price of $100, it generates the same amount: $100 \times \$100 = \$10,000$. However, if it can charge each type of passenger their actual willingness-to-pay, it generates: $50 \times \$200 + 50 \times \$100 = \$15,000$—an increase of 50 percent.

Of course, the big assumption is that the airline can figure out who's willing to pay what and then charge each person accordingly. However, perfection isn't necessary, and it isn't the goal. Capturing even a small fraction of the additional revenue would be a windfall for an industry that operates on margins of a few percent in the best of times.

For airlines, the ability to distinguish between business and leisure passengers can be found in the flying habits of these two groups. Business travelers tend to fly during the work week, frequently booking only a few days prior to a scheduled meeting. Leisure travelers are willing to book well in advance of their travel dates, and will often stay over a weekend. Recognizing this, airline marketing departments went to work designing fare products to capture these various price sensitivities. Fare products blossomed—twenty-one, forteen, and seven day advanced purchase fares, weekend fares, Tuesday fares, flight-specific fares, fares with and without penalties—all aimed at increasing price-sensitive leisure traffic while keeping price-insensitive business travelers paying as much as possible.

Retailers have long understood that there are different market segments willing to pay different prices for a good. A walk through any appliance store where there are dozens of models of refrigerators, washing machines, and microwave ovens makes this vividly clear. Refrigerators differ by size, energy consumption, and convenience features such as ice dispensers. Prices for these different refrigerators may vary from a few hundred dollars to well into the thousands. Though at times we may be overwhelmed, we recognize that we're fortunate to have so many options to choose from. Perhaps paying an extra $100 may seem a bit steep to get the clear plastic shelving instead of the white, or the curved door handle instead of the straight, but at least we're willing to concede that the store has a right to charge a different price for each of the models because they're clearly different products.

When products aren't perceived as different, it's another story. Imagine two people purchasing identical refrigerators at the same store on the same day, one for $500 and the other for $1,000. If the person paying $1,000 found out about the person who paid $500, he would be incensed. In all likelihood, he would make his way back to the store and demand the repayment of $500 as a matter of fairness. A report issued by the Annenberg Public Policy Center at the University of Pennsylvania in June of 2005 makes this point all too clear. Of the 1,500 people polled in a telephone survey, 87 percent disagreed with the statement "It's okay if an online store I use charges people different prices for the same products during the same hour."

Unfortunately, the line that distinguishes different products is not always clear. Delivery services are no longer limited to overnight delivery. We have many different options for when our package will arrive—the next morning, the next day by the close of business, two days, one week—and each option carries a different price. Delivery services have created different products from the same basic activity (delivering a package) by incorporating time in the product definition. One week delivery is a $5 product; overnight delivery is a $15 product.

It could be argued that delivering the package constitutes the entire product and that time-to-delivery shouldn't affect price. Certainly, time-to-delivery isn't a physical attribute in the same way door handles on a refrigerator are. However, we'd almost all agree that it's perfectly reasonable for a company to charge different prices for different delivery times. We place a real value on receiving things earlier rather than later, and we understand that the costs may be different. Incorporating time-to-delivery in the product definition is perfectly acceptable.

Fare classes were originally conceived by airlines as true products, or more exactly, as different models of the same product. Arriving at the store, a passenger looking over the shelves might want the top-of-the-line model (a fully refundable ticket), or he might be interested in something more basic (a nonrefundable ticket). The problem is that beyond refundability, all the other characteristics defining fare products have almost nothing to do with bringing value to the customer. Taking a trip over a weekend and faced with the choice of a nonrefundable ticket priced at $200 or a nonrefundable ticket requiring a Saturday night stay priced at $100, there's nothing about the $200 ticket that would tempt us to purchase it. Fare classes, while viewed as products, were designed with one thing in mind: fencing the population into different segments based on their willingness-to-pay. In the public eye, airlines are charging different prices for the same model of refrigerator.

It's possible to charge different prices for identical items without raising the ire of customers. It happens all the time, though we rarely think of it as differential pricing. Movie theatres offer discounts for children and senior citizens. They attend the same movies at the same time as those paying regular admission, and there are no restrictions, such as where they can sit in the theatre. Because they are receiving discounts as opposed to being charged a different price, and because the discounts adhere to social norms, we rarely, if ever, object.

When airlines began offering different fare products, these products were treated as discounts off regular fares and were marketed as such. But over time, the discount designation disappeared into the background, furthering the perception that airlines are selling the same thing for different prices. One notable exception is Southwest Airlines, which, in addition to offering "refundable anytime" fares (Southwest's term for "full" fares), offers "special fares," "promotional fares," and "fun fares," among others.

Often overlooked is the fact that charging different prices benefits both the passengers and the airline. Consider a slight modification of our earlier example. Now the cost of high-priced tickets is $300, and 50 passengers are willing to pay $300, while another 50 are only willing to pay $100. Forced to set a single price, the airline would choose a price of $300 since 50 × $300 = $15,000 is more than 100 × $100 = $10,000. On an aircraft with 100 seats, this price would cause 50 seats to go empty even though 50 people are willing to pay for them. Both the airlines and passengers lose, even though the airline has made the most profitable decision under the restriction that it can only set one price.

As senior vice president in charge of all planning, scheduling, and operational pricing activities at Continental Airlines, Bill Brunger understands this example as well as anyone. At a conference in May of 2005, Brunger stepped forward with his own example. Examining a tariff sheet issued by the Civil Aeronautics Board in 1980, he observed that the price for a one-way ticket from Newark to Los Angeles on Continental was $334. Twenty-five years later, fare class prices ranged from $329 to $499. For those willing to live with restrictions, charging different prices led to a decrease in the price of a ticket over the 25 year period, even without accounting for inflation.

Brunger went further, however, pointing out that in 1980, the price of a loaf of bread was 48 cents, and the average price of a new car was $5418. In 1980 dollars, the cost to fly from Newark to Los Angeles in 2005 ranged from $130 to $196. Accounting for inflation, even the most expensive price in 2005 was lower than the price in 1980.

So if airlines are doing such a great job of pricing, why have prices dropped? Competition that followed in the wake of deregulation was certainly a key factor. It forced airlines to provide their services more efficiently and produced tremendous cost savings from operational improvements, many of them science-based. Technological improvements in aircraft design and the widespread use of computers also made their mark. But the first decade of the twenty-first century has seen the emergence of two powerful forces putting downward pressure on prices: a new, aggressive breed of airline and the Internet.

We regularly read about low-cost carriers and their battle with the long-established legacy carriers. The low-cost model is straightforward: keep costs down so that prices can be kept low. New, low-cost entrants fly through less expensive airports, offer reduced in-flight service, and benefit from far more favorable labor contracts than established carriers. With cost structures that are well below those of the legacy carriers, the fares offered by low-cost carriers are correspondingly lower than those historically offered by their counterparts. Not surprisingly, in response to these new entrants, legacy carriers have dropped prices to remain competitive.

Less visible is the effect of the Internet. Airlines raced to embrace this new way of selling tickets because it was cheap—far cheaper than any other distribution channel. There was no need for reservations agents to handle the phone, or travel agents to collect fees—The Internet not only reduced costs, but it also helped free airlines from the power wielded by the vast distribution network that had grown up around them. In a presentation to Princeton University in April of 2003, the senior vice president of marketing and planning at U.S. Airways shared what the carrier spent to sell a ticket through various distribution channels: $26 for traditional airline-to-consumer sales such as calls to an airline reservations agent, $21 for third party Internet agencies, $19 for traditional travel agencies, and $11 for the carrier's own Web site, www.usairways.com.

For consumers, the Internet created unheard-of transparency into ticket prices. Rather than calling a travel agent with the hope of receiving the best price available, passengers could compare prices on their own with the touch of a button. Once again, fare class distinctions were further muddled as travel Web sites designed their search engines with a focus on price and schedule. Brunger, who at one time sat on the board of directors at Orbitz, described the design of the Orbitz Web site as an effort to combat a complete emphasis on price. Displaying the lowest price for many airlines at once, Orbitz's Web site allows visitors to see if American Airlines is only a few dollars more than United Airlines rather than simply displaying only United's fare. Other Web site designs would have left American off the list altogether.

Many low-cost carriers adopted a strategy of using the Internet as their primary distribution channel, completely bypassing the historical methods of selling tickets to customers. In an effort to keep costs low, many low-cost carriers don't even work with third party Internet channels, accepting only electronic bookings through their own Web sites.

All airlines recognize that these new and influential forces affecting price must be dealt with, and they are using science to help address them. Experience has shown that for legacy carriers, properly pricing long-distance traffic that connects through hubs becomes increasingly important in the presence of low-cost carriers, who typically focus on point-to-point, short-haul flights. As a result, many carriers are in the process of upgrading their pricing capabilities even as they lose money. Low-cost carriers need science-based pricing, too. While they typically use a simplified product structure, in some cases offering just a single fare class (measured by restrictions), they still vary price in the same way as legacy carriers.

Yet while airlines continue the fight on the revenue front, they're aware that revenue is only one side of the equation. Rising fuel prices have impacted all airlines, but the legacy carriers are also fighting bloated cost structures that can still trace their roots to the era of regulation.

Speaking in April of 2005 at the Georgia Institute of Technology, Brunger lightheartedly weighed in with his thoughts on the state of the airline industry. "If I'd brought a PowerPoint slide," said Brunger as he proceeded to the large, white projection screen at the front of the lecture hall, "it would have looked like this." Facing the crowd, Brunger turned to his right and spread his arms as wide as he could, one hand toward the ceiling, the other to the floor. "This is how much it costs to run an airline." He then turned to his left, again spreading his arms from ceiling to floor, but not quite as wide. "And this," he explained, "is how much revenue we're generating." Smiling, he looked at the audience face-on. "I have a theory, and it's only a theory," he continued, "that *this* bar," arms to the right, "needs to be *smaller* than *this* bar," arms to the left, "for an airline to survive."

Point taken.

* * *

The nation of Papua New Guinea occupies the eastern half of the large island of New Guinea and a collection of smaller islands to the east. Located just north of Australia and just south of the equator, it's a lush, tropical land so densely forested that the many tribal communities who live there exist in relative isolation. While the government consists of a constitutional monarchy with a parliamentary democracy—a byproduct of European influence—the population of 5.5 million people speaks over 700 languages, and on any given day it's not entirely clear who the government is actually governing.

The largest city in Papua New Guinea is the capital of Port Moresby. Home to 350 thousand, there are no significant roads leading beyond the city borders.

The main source of transportation is the airplane, connecting 571 airports, 21 with paved runways.

On a typical hot, humid day, PROS's co-founder Salter found himself standing in Air New Guinea's corporate headquarters. A comfortable, low-rise building set on a bluff, it overlooks the buildings and runways that make up Jackson's Airport in Port Moresby. Engaging in small talk with company executives, he was treated to the story of an event that occurred only weeks before just outside the conference room window.

An executive meeting had been interrupted by the sound of gunfire. A plane carrying pay for the airline's employees had landed, and was being attacked by a group of armed men in a truck. The pilot had been pulled from the plane and thrown to the ground. He wasn't moving.

The armed men quickly made their way onto the plane and found what they were looking for. Jumping back in the truck, they drove to the far side of the airport, crashing through a fence and into the woods. While all of this was occurring, the pilot got back on his feet. But rather than seeking assistance, he ran to a nearby helicopter. Climbing in, he proceeded to chase the men into the woods. Gunfire was heard for many minutes during the helicopter's pursuit. The executives would later learn that the pilot had been able to get help from police on the ground as he communicated the men's location, and the money had been recovered.

Salter was impressed by the pilot's sense of duty, and said as much. His hosts thanked him, but pointed out what in retrospect should have been obvious: the pilot's money was in the truck just like everyone else's.

Not every airline was dealing with the same problems as Air New Guinea. Some of the larger commercial carriers in the United States and Europe were already developing a solid scientific foundation for operational pricing by the mid-1980s. Their success and their willingness to share information about that success undoubtedly helped kindle interest for pricing software at airlines like Air New Guinea. But the industry as a whole was grappling with many of the most basic pricing issues.

The question facing Woestemeyer, Salter, and Yuen was how to bring about change. The motivation wasn't entirely altruistic, since the objective was to sell software and build a business. But this would only occur if they could convince potential buyers of the value of their wares. At first glance the task seemed easy. The financial benefit was apparent and airlines were increasingly aware of the damage a bad inventory controller could inflict. Yet this was a fundamentally new kind of software. It didn't automate an existing business practice like paying bills or filling out invoices. Nor were the purchasers engineers who would willingly adopt the latest technology. If anything, they were some of the less technical people in the airlines, people who had learned to rely on their gut instinct. All these issues combined to make for a difficult sales process.

No answer immediately presented itself, and although Woestemeyer, Salter, and Yuen learned as they went, they quickly realized there was no single solution. Every organization had its own unique character and its own set of problems. The situation was further complicated by the fact that they were dealing with cultures scattered around the globe. Business in South Africa is very different than business in China or Saudi Arabia.

Yuen, who works with clients throughout the world, believes that establishing a relationship of trust is one of the most important factors in selling a system. "You must show that you understand their business and that you're willing to do what it takes to support them as people," said Yuen. "And once you've earned that trust, you need to work hard to keep it."

Yuen understands this more than anyone. When he scheduled a Christmas vacation in Italy one year, word spread quickly through the office. Yuen simply didn't take vacations, and the consensus was that it was long overdue and certainly well earned. Arriving at his hotel, he found a fax waiting from an important client in Asia that he'd been working with for years, and that he was hoping to sign a contract with shortly.

"I spent the majority of my vacation in the hotel, rounding up people at the office in Houston and working the fax machine," said Yuen. "The client had pages of detailed questions, and they wanted the answers by December 26." The situation could have been written off as a cultural misunderstanding from a country that doesn't celebrate Christian holidays, but Yuen understood that it was more.

"It was a test," said Yuen. "They wanted to know if we were willing to work during the Christmas holiday." Did he mind? "No, not really. It's just part of what you do."

Relationship building went far beyond tests. On one occasion, a team visiting an Asian client received word that a very senior executive of the airline would be coming by to offer his greetings. It was a fortuitous and unexpected meeting, and Salter was convinced by the team that his presence would be seen as a sign of commitment. Shoulder bag in hand, Salter hopped on a plane and flew halfway around the world for a meeting that lasted less than 30 minutes. Finished, he boarded the next available flight for his return to Houston.

Relationships, however, didn't get to the heart of the value proposition, which was the science. Woestemeyer, Salter, and Yuen didn't have scientific backgrounds, but each understood its importance when making a sale.

Yuen focused on what he calls the "theory of one." Be it one more full fare passenger per flight or a 1 percent increase in revenue, one seemed to be a number few people could argue with. "We usually started with bigger numbers and rounded down to one," said Yuen. Getting five passengers on a 130-seat aircraft to pay an extra $200 doesn't seem unreasonable. So suppose this number is rounded down to one passenger paying an extra $100. When you multiply that by the total departures in a day, the numbers add up quickly. Even with these very conservative estimates, Yuen could show that the systems paid for themselves in a matter of weeks or months.

Woestemeyer was always on the lookout for "Tom," his generic name for the brightest guy at an airline doing inventory control. "If you could convince Tom," said Woestemeyer, "everything else fell into place." But Tom could also be an entrepreneur's worst nightmare. Someone who's convinced that he's right and you're wrong can easily find himself stepping over the line to prove his point. At one carrier, a particularly strong-willed analyst was certain he could overbook flights better than a piece of software. Woestemeyer had already sold the system and installed it at the carrier, but the analyst persisted. When it first came online, the proposed overbooking limits on a handful of flights were so high they were clearly erroneous, primarily due to inadequate historical data. Woestemeyer and the entire team were aware of these cases and coached the analysts to limit overbooking levels until the system had a chance to learn. On a busy holiday weekend, Tom intentionally ignored their advice.

Too much overbooking doesn't make anyone happy. Passengers are rightfully angry, gate agents are pushed to their wits' end and let management know about it, and management is furious at the impact on customer relations. Woestemeyer managed his way through the situation because what had occurred was obvious, but it was an unpleasant experience. Weeks later, as the results of the system were measured, the impact became clear. Load factors—the number of seats filled on departing planes—had jumped 8 percent with no measurable increase in involuntary denied boardings. The revenue implications were staggering.

Like all of us, the analysts in charge of overbooking were risk averse. When operating without the help of a system, on the rare occasion when they would overbook a flight to a point at which passengers became angry, they would have to deal with irate gate agents. Analysts don't like unnecessary conflict any more than the rest of us, especially when we feel at fault. As a result, we play too conservatively and draw upon our bad experiences to defend our actions, often under the guise of common sense.

The common sense versus science debate was a major theme as Woestemeyer, Salter, and Yuen sought to build a market for scientific pricing. For many carriers, just getting them to understand the problem and what science could provide was a significant obstacle. All three men point to examples of sales efforts that took five, six, or seven years—a remarkably long time for a relatively inexpensive system. It took time for the carriers to become comfortable with the idea of letting software choose what price to make available, and many were willing to sit and wait to see what their competitors did. Yet as Woestemeyer's experience aptly punctuates, even those carriers who saw the need for a better way to price often had trouble convincing the frontline workforce.

* * *

Woestemeyer was born and bred in the Lone Star State, though you wouldn't know it to meet him. With no accent and no special attraction to things cowboy, he belongs to a new breed of Texan. He hasn't, however, entirely escaped the practice of storytelling to make a point. "Pricing is like a frog in a pot," he told me. "If you put 'em in cold water and turn up the heat, they'll die before they figure out what's happening." I asked what this had to do with pricing. "A lot of companies are frogs, and scientific pricing's the heat."

Woestemeyer started Propsys in 1983 in a small, two-story building in midtown Houston, just south of downtown. In Texas tradition, the boundaries of midtown and downtown are defined by freeways. Propsys would later become PROS, an acronym that has represented different things at different times, but would eventually stand for Pricing and Revenue Optimization Solutions.

As people began moving back into the city to live, the office would be converted into a fashionable townhouse. But in the 1980s, the location was far from spectacular, even though it was only a quarter mile from the many high-rise office buildings in downtown. Police raids at the "modeling" studio located next door were a common occurrence. Thanks to Houston's nonexistent zoning, the office was situated directly across the street from what came to be known as the "exotic zoo." Llamas were observed on the property as cars came and went, but the fences were so overgrown that no one was ever entirely sure what animals resided there. When the wind blew from the west, the smell could be overpowering.

Generating capital to grow the company, Woestemeyer and his wife Mariette Melchior—a summa cum laude graduate of the University of Houston—would at one point find everything they owned tied up in the business. "Looking back, it was a huge risk," confided Melchior. "But we didn't have any children and kind of thought, 'What the heck?'" Melchior, who can be tough as nails when circumstances require, earned money by helping her sister open a bridal shop in their home country of Luxembourg. At one time, she handled shipping and receiving through the PROS office in Houston. "All the boxing and taping used to drive our programmer crazy when he was trying to work," she said with a grin that betrayed just how much she enjoyed that part of the job.

Selling science required knowing science, and for this, Woestemeyer and Salter turned to the academic community, a business strategy the company has embraced throughout PROS's lifetime. Salter had his first contact with the science community at Continental when Lorenzo gave him a call. Lorenzo, who completed an MBA at the Harvard Business School, had heard of a student who had coauthored a paper in the Journal of Transport Economics and Policy on the price elasticity of demand for air travel. The work sounded like it might have the potential to provide Continental with a competitive advantage. Salter contacted the student and hired him to work at Continental. Unfortunately, the airline wasn't quite ready for the young academic and he left after a few years to work for his father in Michigan. Woestemeyer would later hire him away from his father to help get PROS off the ground, but the death of the young man's father precipitated a return to Michigan after only a brief time at PROS. Shortly thereafter, Woestemeyer connected with a PhD student named Peter Belobaba from the department of aeronautics and astronautics at the Massachusetts Institute of Technology. Belobaba completed one of the first doctoral dissertations in revenue management and would play an important role in shaping PROS's scientific direction for many years.

Early on, the science was relatively straightforward. Computers had plenty of data but they weren't good at providing it in a way that could be used. Elaborate mathematics took a backseat to more basic problems like managing data and using it to set reasonable inventory levels for the many flights each carrier operated. As these problems became less of an issue, the door was opened to more unrestricted thinking about revenue management. Were there better ways of forecasting? What about better algorithms for setting inventory levels? No one had a road map, but the options seemed limitless.

Responding to the market, companies selling revenue management software or consulting services began using science as a means of competitive differentiation. The skills required to write software, manage projects, and introduce new business processes weren't valued by airlines in the same way that science was, even though time would prove that these skills were essential to a carrier's success. Selling science was no longer missionary work, especially for the more lucrative contracts. It was a struggle for the intellectual high ground.

By the early 1990s, PROS faced three major competitors: Aeronomics, DFI, and SABRE. Aeronomics was started by a former executive at Delta who had been through many of the same experiences as Woestemeyer, Salter, and Yuen. DFI was founded by a group out of Stanford University that built its reputation by hiring people with advanced degrees from top schools. Both companies were primarily consulting organizations, though they did write software. Among other clients, DFI had worked with United and Scandinavian. Both airlines were highly respected for their revenue management efforts, which were considered leading edge at the time.

The SABRE name was closely tied to American Airlines. Admired and feared by its competitors throughout the late twentieth century, American recognized the importance of science in everything from pricing to schedule planning. In 1986, SABRE became a fully formed division of AMR, the holding company for American Airlines, which gave it unrestricted freedom to market its products, services, and expertise to other airlines. Over time, the scientific arm of SABRE would go through a sequence of name changes.

Each competitor had its relative strengths and weaknesses. SABRE was by far the largest and offered a suite of products to complement its revenue management software. It also had a long history of applying science to airline problems. But being a former arm of the ferociously competitive American Airlines was both a strength and a weakness for the company. With its hiring practices and the way it marketed itself, DFI was considered a boutique company offering the very best in pricing science as it helped companies design and build custom systems. Aeronomics had a history of science as well, going so far as to publish a respected newsletter filled with as many equations as articles. Even so, the company's strength was generally viewed as its business acumen.

Although PROS had developed a solid science underpinning through its work with academic experts and its practice of hiring bright, young people, it didn't have a reputation as a science leader relative to its competitors. Though it maintained contact with the academic community, by 1997, its internal research and design group, which had never grown very large to begin with, had fallen to one person. Yet PROS did have far and away the largest number of clients.

The elements that contributed to PROS's success were many. Relationships were certainly part of the story, and PROS had tremendous depth of expertise in airline revenue management. It was also an incredibly scrappy, independent-minded company. Woestemeyer and Salter wouldn't share details, but there were heavy-handed efforts by more than one large corporation seeking to buy PROS under less than desirable terms. In each case, the interaction only furthered the resolve of everyone involved with PROS.

A focus on software products instead of consulting also helped the company. Although it would struggle throughout its existence with the balance between maintaining a standard product and satisfying client requirements, PROS always kept a product mindset that ultimately helped everyone. Standard products could be updated and improved. They become part of the collective knowledge of the company rather than of the last person who worked on them, thus reducing risk for anyone dependent on the software. As the market would prove, these benefits almost always outweighed the advantages of custom-built systems.

One of the often overlooked elements contributing to PROS's success was that the early science it embraced wasn't overly complicated. Most of the science could be explained to clients so that they understood it. By incorporating straightforward methodologies, users could check results and grow comfortable with the system. More basic mathematical methods were also less temperamental than their advanced counterparts and made it easier to diagnose problems when they occurred.

The more basic methods also provide a large fraction of the potential revenues. Though the figures vary from carrier to carrier and person to person, practitioners in the airline industry commonly quote revenue improvements in the neighborhood of 6 percent when using a good but basic revenue management system. Moving from basic revenue management to more complicated methods can yield up to an additional 3 percent revenue increase. For most organizations, a 3 percent improvement is a lot of money, and is well worth pursuing. But when an industry is just coming to terms with science and its organizational impact, a basic approach is usually the right place to start.

From their earliest sales efforts, Woestemeyer, Salter, Yuen, and their competitors had focused on a problem of vital importance to airlines—filling seats profitably. Although it took time and effort to make their voices heard, they helped bring a level of mathematical sophistication to operational pricing that had never before been experienced in any industry. Without question, method had come to a world in dire need of it. But had they eliminated madness, or simply supplanted it with something new? Answering this question requires a further look into the details of airline pricing.

Notes

- The award-winning treatise referred to in the text is Talluri and van Ryzin (2004).
- Statistics on Continental Airlines are taken from the company's 2004 annual report.
- The presentations given by Bill Brunger took place at the 2005 PROS Revenue Management Conference, Houston, TX, and the Georgia Institute of Technology Revenue Management and Price Optimization Conference, Atlanta, GA.
- The Princeton University presentation was made by Ben Baldanza.
- Information on Papua New Guinea can be found at https://www.cia.gov/cia/publications/factbook/geos/pp.html.

CHAPTER 3

The Computer Did It

S cientific pricing wasn't something airlines envisioned as commercial aviation took shape in the 1920s. Far from it. Eddie Rickenbacker, a celebrated World War I flying ace, raced cars before purchasing and running Eastern Airlines in 1938. Cyrus Rowland Smith, an early president of American Airlines, was known simply as "C. R.," the hard drinking salesman from Texas. These certainly weren't people you'd expect to scour universities looking for the world's brightest pricing scientists.

What happened was data. Airlines need to be careful about record keeping. Every time a seat is booked, information must be kept about the passengers and flights that are involved. Passengers may want to call and change their plans. If a schedule is changed, passengers need to be contacted. The number of seats sold on each flight must be tallied so that planes aren't oversold (or, at least not by too much). This process of record keeping served as the catalyst for a remarkable sequence of events that would push computer manufacturers into completely new realms, lead to the earliest demonstration of e-commerce, and ultimately pave the way for science to establish itself as an essential business tool.

Passenger information was originally recorded in ledgers. Upon receiving a request to purchase a ticket, a reservations agent would pull from a shelf the ledger containing information on the requested flight. If the flight had seats available, the agent would write down the passenger's name and contact information. If the passenger wanted to cancel the reservation, the information was erased. The term *booking* is still used to refer to the process of recording a reserved seat, and it is even more prevalent than the word *reservation*.

Ledgers simply couldn't keep up with the growth of commercial aviation. A company owning 10 aircraft might operate a schedule of only 20 daily flights. But bookings are taken in advance of the departure date. If reservations are accepted 30 days prior to a flight's departure, then at any given time, agents are managing some 30 instances of the 20 flights, or some 600 flights in total. With 25 seats per plane, the approximate number aboard a Douglas DC-3, agents at the 10 plane airline would find themselves managing 15 thousand seats.

Today, the numbers are magnitudes larger. In 2004, American Airlines operated 1013 aircraft with an average seating capacity of 130. Assuming two flight departures per day for each aircraft and a window of 360 days prior to departure during which a passenger can book a flight, on any given day, American is managing an inventory of close to 100 million seats.

Airlines were acutely aware of the numbers problem from very early on. Ledgers quickly gave way to three-by-five index cards kept in boxes or file drawers. Simple as this change seems, it offered many advantages. Cancelled reservations were no longer erased, they were thrown away. Flight changes were handled by moving a card from one box to another. For many years, reservations agents at American filed colored cards in boxes atop swiveling circular platforms situated on tables—an industrial version of the Lazy Susan. Because the colors were reminiscent of the popular Tiffany lamps of the era, this method of managing reservations was informally referred to as the "Tiffany system."

To make a booking, however, an agent still needed to know whether a flight had any seats available, which required tracking down the appropriate box. Recognizing this problem, airlines tried various ways of publicly posting flight availability. The first attempts consisted of writing flight numbers on chalkboards with basic information such as "seats available" or "seats not available." Faced with a call requesting a seat, a reservations agent could look at the appropriate chalkboard. If a seat was available, the reservations agent would write the necessary information on a card and have it delivered to a person who kept records on the flight. In turn, that person would check to see if the new booking changed the status of the flight and, if so, would update the chalkboard.

Decades after the system of chalkboards was introduced, Al Ludwig began his career with the airline industry. Fresh out of high school and looking to make enough money for college, a friend's father helped Ludwig secure a job as a reservations clerk at TWA's regional office in Pittsburgh. His job was to update the boards showing publicly posted flight availability. However, Ludwig didn't carry a piece of chalk and an eraser. Instead, he had at his disposal many brightly colored chips.

The boards had gone through technological advances. Similar to the style used in large university lecture halls, the boards stood two stories high and used a system of pulleys so they could be moved up and down the wall. Thirty days prior to departure, a flight's number would be added to the board next to a nail on which a two-inch chip could be placed. No chip meant a flight had seats available, while a green chip indicated a flight was full but passengers could be placed on a waiting list. A red chip meant the waiting list was closed. Ludwig was issued a long, specially designed stick that could be used to replace chips that were out of his reach.

Keeping the chips up to date could be a difficult task if too many flights changed status at the same time. When Ludwig's manager got wind of a discrepancy between the actual bookings and the status of the board, the manager would march to the front of the room, pull down the appropriate board, and huffily make the change himself. As Ludwig recalls, on more than one occasion the manager would slam the board down a bit too hard, causing all the chips to fall into a trough at the bottom. Until records were checked and the chips replaced, reservations agents were unable to make bookings for flights on the empty board.

The boards and their chips stood at the front of the room where reservations were taken. The room contained rows of desks lined up in parallel lines, much like dinner tables in a camp mess hall. Reservations agents sat along one side of each row facing the front of the room and the boards. On the other side of each row was a conveyor belt. When an agent concluded a call, she (almost all the agents were women) would place her three-by-five index card on the belt. The conveyor belts ran to one side of the room where they made a 90 degree turn onto another belt that gathered the cards and sent them to a room where they would be collected and filed. The conveyor belts were also a technological advance, having replaced card runners from years earlier.

Constantly operating mechanical devices that they were, the belts would regularly break. There was no scheduled maintenance for the belts, nor was there a special group to handle repairs. When a belt broke, floor supervisors and reservations clerks worked with whatever tools were available to fix the problem. Broken belts could be a nuisance, but even more problematic were the tribulations that followed the call "Belt jam!" While making their 90 degree turns at the end of a belt, cards transitioned to the next belt via a spinning wheel. If a card became jammed between the belt and wheel, hundreds of cards could become entangled before someone noticed. Because of the way the cards jammed, they were frequently destroyed in the process. Passengers with lost cards were destined to hear "I'm sorry, but we have no record of your booking" at some later date.

As the rooms grew bigger and bigger, it was clear that an alternative solution had to be found. Reservations agents were keeping field glasses on their desks so they could see the distant boards. The system simply wasn't capable of scaling much further.

The earliest known experiment with a mechanical device for counting passengers dates to the mid-1940s at American Airlines. Akin to the candy dispensers often seen at movie theatres, the machine consisted of a number of tall cylinders, one for each flight. The number of balls in a cylinder corresponded to the number of seats available. When a reservations agent made a booking, a button was pushed, causing a trap door on the appropriate cylinder to open. A ball would fall out, recording the booking. Cancellations were handled by adding balls.

Shortly after this early experiment, American worked with the Teleregister Corporation to devise a slightly more advanced device known as the Magnetronic Reservisor, which it put to work in 1952. It used an early version of a magnetic drum to store essentially the same information as the gumball dispensers. Those who worked with the Reservisor referred to it as "Girlie," since she "told all."

At TWA, Ludwig worked with a later version of the device known generically as the Teleregister Telefile, or simply Teleregister. Inserting a metal card corresponding to the desired flight into a Teleregister, Ludwig would wait for a response that translated into a chip: a solid light corresponded to no chip; a slow flash, to a green chip; and a fast flash, to a red chip. Eventually, a Teleregister would be issued to each reservations agent, thus eliminating the need for boards and chips.

The Teleregister had obvious limitations. It provided a quick way for reservations agents to determine if a flight was full, but information about individual passenger bookings was still recorded on cards and filed. Information was fed into the Teleregister by someone who counted passengers on cards, determined what state the flight was in, and then manually entered that information.

Clearly, there was room for technological improvement. The obvious solution is something we take for granted today: a computer reservations system that allows reservations agents to type passenger information on to a screen once, then push a button that files the information and simultaneously updates passenger counts on flights. The problem was that commercial computers didn't exist.

The IBM 701, released in 1953, was the first commercial computer manufactured by IBM in quantity. By today's standards it was quite primitive, with memory designed around cathode ray tubes. That same year, a chance meeting between an IBM representative and the president of American Airlines, C. R. Smith, would lead to the first computerized reservations system. The project was deemed SABER, for Semi-Automatic Business Environment Research. Joint experimental work was carried out by the two companies, ultimately leading to a contract to build a reservations system in 1959. At the time, IBM was also talking to both Delta and Pan Am about building similar systems, and internally named the umbrella project for all three airlines SABER. As a result, American renamed its system "SABRE," with the "E" and the "R" interchanged. Both spellings are acceptable for the long, heavy military sword that unquestionably inspired the acronym.

SABRE ultimately came on line in 1962 and became fully operational in 1964. Delta's DELTAMATIC and Pan Am's PANAMAC followed a year later in 1965. As the three carriers were completing their installations with IBM, United and TWA began their own projects, United with Univac and TWA with Burroughs. At TWA the system was named "George," as in the phrase, "Let George do it." Coming to the game late, the new players sought to build reservations systems with functionality that would leapfrog their rivals. Unfortunately, by 1968, when the systems were scheduled to be completed, it was growing increasingly clear that they were facing insurmountable problems. At TWA, a young man by the name of Bob Crandall took responsibility for unceremoniously killing George and writing off the phenomenal $75 million that had already been spent on it. Crandall's no-nonsense action in a politically tumultuous environment foreshadowed a style that would lead to his eventual rise to chairman and CEO of American and his reputation as the most feared and respected competitor in the history of the airline industry.

Both United and TWA would recover from their misstep. Based on experience with American, Delta, and Pan Am, IBM would go on to develop PARS, the Programmed Airline Reservations System, which it offered for sale in 1965. Among the purchasers of PARS was Eastern. Eastern expanded and developed the basic PARS system, which was later purchased by both United and TWA. TWA continued to call the system PARS, while United changed the name to Apollo as it further expanded the system's capabilities. Even American purchased the Eastern-based PARS system, completing integration with the SABRE system in 1972. While reservations systems went through further evolution, name changes, and shifting of who owned what, by the late 1970s, they had firmly established themselves as an essential part of operating an airline.

It would be impossible to overstate the impact of reservations systems on the airline industry. With the information about passengers previously held on three-by-five cards now maintained as an electronic record, and with the ability to quickly and accurately link this information to actual flights, mistakes in the booking process were greatly reduced. Chalkboards and colored chips were gone, as were the many people necessary to maintain them. For airlines that made the investment in reservations systems, things were running much more smoothly.

But there was an additional advantage that not all airlines were immediately aware of. Information about how passengers flew was now being gathered in a form that facilitated scientific analysis. By understanding traffic patterns, airlines could develop network flight schedules that best served passengers' needs and, not coincidentally, generate higher revenues for the airline. Airlines could also pick up on changes in flying patterns and modify schedules more rapidly and intelligently. It was also possible to observe exactly how many people were paying for the many different fare classes.

Scientific advances, however, occurred slowly. Early reservations systems were also early generation mainframe computers. Designed to facilitate the reservations process, reservations systems had the data necessary for scientific analysis, but it wasn't always easy to get at. In addition, while the science necessary to analyze the data was developing at a rapid pace, the actual implementation of this science on computers was still in its infancy. But most importantly for airline pricing, reservations systems had to go through another evolutionary stage before their full potential could be realized.

* * *

The internal record keeping capabilities of reservations systems were an important advance for airlines. Selling tickets, however, still required reservations agents to sit by their phones, ready to take orders from potential passengers or their travel agents. The distribution system—the process by which airlines reach the end consumer—had yet to be automated.

For many years, travel agents played a central role in airline ticket distribution, offering advantages to passengers and airlines alike. Working through a travel agent, a passenger could expect to receive a reasonably priced ticket from whatever airline best met his needs. Travel agent offices were often located in the passenger's neighborhood, thus eliminating a trip to an airport or a downtown airline ticket office to complete the purchase. Airlines viewed travel agents as their commissioned sales force, without needing to pay for the overhead of setting up a ticket office and staffing it with salaried employees.

For their part, travel agents dealt with many of the same problems faced by early airline reservations agents, but at a different level. When a passenger called a travel agent for help in booking a ticket, the travel agent would look at printed paper schedules to find reasonable flights, then use the phone or teletype to reach an airline and determine seat availability. The agent could wait minutes or hours for a response, and if the desired flight was full, she would have to repeat the process. What travel agents needed was a way to locate flights and available seats more efficiently. What they needed was a system similar to what the reservations agents at airlines had with their reservations systems. However, the travel agents didn't need flight information for a single airline; they needed flight information for all airlines.

From 1967 to 1974, multiple attempts were made to develop a reservations system that would serve the needs of travel agents. One of the most significant was a joint effort of the American Society of Travel Agents (ASTA) and the Control Data Corporation. In 1973, the two organizations signed an agreement to begin work on their own reservations system for travel agents.

Given the important role of travel agents in the sales process, the airlines were rightfully worried about the power that would be afforded by the new reservations system, if it were ever built. If its purveyors were successful at getting a computer terminal on each travel agent's desk, airlines would be at the mercy of ASTA and the Control Data Corporation and whatever fees they wanted to charge. As a result, American, under Crandall's leadership, took action and convinced a group of airlines, computer makers, and ASTA members to look into developing a broader industry initiative known as the Joint Industry Computerized Reservations System.

For two years, American led an effort to study the feasibility of developing a joint industry solution, and in a report issued in 1975, the company concluded that building such a system was not only possible but also a sound financial decision on the part of the industry. But as with so many cooperative agreements, mistrust and competitive wrangling led the initiative to fall apart. United decided to break ranks and to offer its own powerful Apollo reservations system—with some modifications—to travel agents and large corporate travel offices. Because competitors' flight schedules and fares were already available through industry organizations such as the Official Airline Guide and the Airline Tariff Publishing Company, the offering by United could include information for the entire industry, not just United's flights. In January of 1976, United announced that it would accept orders in May of that same year and begin making deliveries in September.

American had been aware of United's arms-length role throughout the two-year study. Either as a competitive countermeasure or a carefully reasoned strategy from the time the joint industry initiative was conceived, American was ready with its own response. It moved quickly, installing SABRE in some 130 travel agent offices by year's end—far and away more than United. The race was on.

If selling access to their reservations systems was initially the result of a sequence of defensive measures, over time it became apparent that reservations system access was an important source of revenue. No company was more aware of this than American, where Crandall established Travel Agency Automation as a separate business unit. However, the revenue was generated not only from travel agents who paid to use the reservations systems, but also from the sale of more tickets due to screen bias and the halo effect.

Screen bias takes advantage of the natural human tendency of travel agents to do things as simply and expeditiously as possible. Industry studies indicated that approximately 70 percent of all booked flights were taken from the first computer screen presented to a travel agent. While airlines that sold access to their reservations systems provided information on many airlines, there were no guarantees on *how* that information would be displayed. Thus, if a travel agent were seeking a flight departing at nine o'clock in the morning from Dallas to Chicago, a travel agent with a SABRE reservations system might see the eight-thirty and nine-thirty departures on American before the nine o'clock departure on Delta. In fact, there was nothing stopping SABRE from displaying every American flight before displaying any competitor's flights at all. There were no legal restrictions, just the market demands of the paying travel agents. If the airline behind the reservations system went too far, a travel agency could always find another reservations system provider. The halo effect refers to another natural human tendency: a generally positive inclination toward American that resulted in the booking of an American flight simply because American and SABRE were attached in travel agents' minds.

In a comprehensive study completed by the U.S. Department of Transportation, it was estimated that during the period from 1976 to 1986, SABRE generated a cumulative $31 million, calculated by taking fees charged to SABRE system users less expenses to maintain the system. However, if additional revenues taken from competing airlines because of screen bias and the halo effect were accounted for, the net amount generated by SABRE during the same period was somewhere between $220 million and $1.8 billion. While this wide range reflects the fact that calculating the additional revenues resulting from psychological factors is a difficult task, it remains clear that screen bias and the halo effect represented huge sources of income.

Everyone in the industry could see the benefits that a carrier received from having its reservations system in place with a travel agent, and that American and United were the undisputed leaders in capturing market share. In 1983, American and United respectively secured 43 and 27 percent of all revenues booked through reservations systems. TWA, Eastern, and Delta were also in the hunt with their own offerings, but they lagged well behind. With huge start-up costs, carriers without reservations systems couldn't hope to build their own. By 1985, 86 percent of all airline tickets were being sold through travel agencies. As travel agencies continued to adopt reservations systems at a frenetic pace, the carriers without them were quick to realize the financial impact would be devastating.

Fortunately, the intense competition between American and United opened a new avenue for carriers without reservations systems known as *co-hosting*. In a co-hosting arrangement, an airline pays a fee to receive preferential treatment on the displays of travel agents using American's reservations system. In this way, the airline could benefit from screen bias, capturing one of the major benefits of having a reservations system without actually owning one.

While at first glance, this might appear to be at cross-purposes with what American was seeking to do with SABRE, it was actually an act of marketing genius. A carrier competing with United in a particular city might find it was losing market share as United marketed its Apollo reservations system to local travel agents. Without a reservations system to offer, the carrier would have no recourse. By choosing to co-host with SABRE, it could strike back with a competitive offering. This action helped the carrier, but it also helped American, which received fees for the co-hosting arrangement, while simultaneously gaining a foothold for SABRE in a new city. And, of course, anything that put a competitor like United in a defensive position was always a bonus. Soon, co-hosting became a standard practice for all airlines that owned a reservations system.

The large carriers with reservations systems could also use co-hosting as an enticement to regional carriers. As their name suggests, regional carriers fly smaller networks and are important sources of passengers for the larger carriers. By offering favorable co-hosting arrangements, a larger carrier with a reservations system could help increase its own traffic while simultaneously increasing the traffic of the regional carrier. The arrangement could be so advantageous for the larger carrier that at times it led to competition between large carriers for a regional carrier's business. Serving the city of Dallas, home to American and Braniff International Airways as they fought for turf in the late 1970s and early 1980s, the regional carrier Texas International was the beneficiary not only of co-hosting, but also paid promotions, print advertising, collateral material, and even television time; this is not to mention free first-class tickets, first-class hotels, and rental cars for its executives.

Reservations systems became such a dominant part of the competitive landscape that in 1982, the U.S. Department of Justice began a preliminary investigation to determine if airlines were using them to the detriment of consumers by promoting anticompetitive practices. Together with the Civil Aeronautics Board, a study was completed that led to a set of rules that would eliminate screen bias, establish a level playing field for the fees charged by reservations systems providers, and generally promote competition. The effort led the Civil Aeronautics Board to put in place formal restrictions on the use of reservations systems, restrictions that would be expanded in later years by the Department of Transportation. Airlines were no longer regulated, but reservations systems were.

The rules did not go far enough to satisfy the airlines that competed with American and United, and in 1984, a group of eleven airlines filed suit against the carriers, seeking $250 million from American and another $150 million from United. American and United, which had invested years of effort and hundreds of millions of dollars in their reservations systems, were not quick to back down. Thus began many turbulent years of litigation between the reservations system haves and have-nots, years that would reach an ironic and unforeseen conclusion.

In 2004, the federal government ceased regulating reservations systems. Once again their owners were free to practice screen bias, set fee structures, and display whatever flights they wanted to. The change wasn't due to a shift in government policy, however, but rather to a change in airline distribution.

Two major factors led the federal government to reach its decision. First, the major reservations systems were largely divested from the airlines that built them. With responsibility to a new set of shareholders, those who operated reservations systems were interested in their own financial future, not that of the mother airline. With the divestiture came a slow shift from the name *reservations system* to the now more common *global distribution system*. The name *reservations system* is still commonly used for the internal systems maintained by airlines to manage their own bookings.

Second, the Internet does what reservations systems do, only better. Apart from the arcane commands, sitting at the desk of a travel agent connected to a reservations system isn't that different from sitting at your home computer and linking to a Web site that sells tickets. More importantly from a regulatory perspective, the cost of setting up and selling on the Internet is virtually nothing compared to the cost incurred by the early airlines who built, sold, and maintained proprietary electronic distribution networks surrounding their reservations systems. The anticompetitive threat once posed by the large reservations systems has largely disappeared, and the long-term viability of today's global distribution systems is now in question.

The reservations systems wars served as an incredible learning opportunity for all those involved with the airline industry. Initially conceived as a means to simplify the process of maintaining internal records on flights and passengers, when faced with the threat of being held hostage to a system built by travel agents, airlines fully realized the power of reservations systems. By the 1980s, the airline industry understood what many industries would wait years to discover: the computer isn't just a record-keeping device, but a competitive weapon in the battle for the consumer. Those who couldn't understand how to play in this new world of computer-based sales and distribution would be severely disadvantaged or go out of business.

As airlines grew to understand the importance of computers in their frontline campaigns for the consumer, there was another fight being waged deep within the airlines themselves. Reservations systems were gathering sales data at a rate and level of detail never before seen in the history of the industry. In the process of selling seats, airlines were maintaining records not just of the number of seats sold on each plane, but also the price and flight itinerary of every passenger flown. The data was available to make intelligent, competitive pricing decisions, but there were organizational and technical issues to deal with.

* * *

Before joining Woestemeyer at Propsys, Robert Salter spent the years from 1979 to 1986 as vice president of sales at Continental Airlines. Fresh out of the deregulatory world, he dealt firsthand with the changes taking place in pricing and inventory control. Within airlines, the sales department is typically responsible for maintaining relationships with large purchasing groups, such as travel agencies, corporate accounts, and wholesale discounters. At Continental, sales was also responsible for airline revenues along with marketing. The pricing and scheduling departments reported to a separate planning group, with pricing creating the fare products and setting prices, and scheduling determining which markets and routes the airline would fly. "We were responsible for how much money the airline made," recalled Salter. "But we didn't have control over products or prices. There were times when we could sell like crazy and the airline would still lose money."

To make matters worse, a group known as Consolidated Reservations Control, or CRC, reported to the reservations department. Operating under the order-taker model, reservations was a service organization much like flight attendants and baggage handlers. "CRC actually managed inventory levels," said Salter. He continued,

> A supervisor, who was a union employee, had the authority to open and close fare classes to allow more bookings if he had an irate passenger or to clear a waiting list. There was no internal system that tracked this, and because the process was fluid the CRC supervisor had enormous control. If you wonder if this control was ever abused, the answer is all the time by every director and above at the airline. Everybody had a friend or a big customer that they needed to take care of. It got so bad that I instituted a policy that no fare class booking limits could be changed without my approval, period.

What Salter realized, as did people with similar responsibilities at other airlines, was that the role of reservations had changed. The inventory controllers may have felt that they were doing a good job, keeping management happy and the planes full, but in reality they were feeding an industry that seemed intent on diluting its own revenues. If getting a lower fare was as easy as asking for it, why pay more?

For years Salter battled to transform reservations from a service organization to the sales business unit it needed to be. Inventory control was only part of the problem. Routes that had no competition under regulation might now have one or more competitors offering different prices and different schedules. It was vital for reservations agents handling phone calls to book passengers immediately rather than risk having them call another airline. Salter's efforts to bring about change were hampered by the fact that the reservations agents were represented by the Teamsters, necessitating negotiations for almost anything he wanted to implement.

One of the changes Salter championed was to place reservations agents on commission. In an effort to promote competitiveness, he set aside a wall some 30 feet in length showing the performance of each of the nearly 400 reservations agents in the Houston office by name. By looking at the wall, a reservations agent could immediately see how she was doing relative to all the other reservations agents. In addition to recognizing strong performers, this helped encourage weak performers to look elsewhere for employment. While such practices are common in sales environments, it was a radical change for the unionized reservations agents.

In the early 1980s, reservations agents made around $28,000 per year. Salter's goal was to pay at least one reservations agent a total of over $50,000, with the amount above salary coming from commissions. The increase in productivity required to reach this goal would serve as proof that implementing commissions for reservations agents was a good idea.

With the commission plan in place, it was not long before Salter found his poster child. A bright, ambitious young man in his early twenties was well on his way to making Salter's goal. Many of the reservations agents were showing improved performance under the commission program, but the young man was far and away the most outstanding performer. Salter couldn't have been happier.

The only problem was that his performance seemed a little too good, and indeed, it was. During off hours, the young man was removing the names of bookings made by other agents and replacing them with his. Not only was he improving his own performance, but he was doing so at the expense of his coworkers. Faced with no other choice, Salter let the young man go. "It hurt," said Salter. "This kid was young and smart and would have been a great success. He didn't need to steal, but he couldn't deal with just being good. He wanted to be the best."

The young man wasn't alone in his efforts to abuse the commission program. A reservations agent who found herself a few bookings shy of the daily quota could simply make fictitious bookings. The problem first surfaced when calling telephone numbers attached to the alleged bookings. If the number didn't exist, or if the person on the end of the line had never heard of John Doe, it was a pretty good sign that something was amiss. Salter fixed the problem by making arrangements to gather post-departure information, and agent's commissions were henceforth based on booked-and-flown passengers, not booked passengers.

Beyond inflated commissions, fictitious bookings had the serious and undesirable side effect of artificially inflating passenger no-show rates. Planes that appeared fully booked might fly with empty seats because a reservations agent had purposely filled those seats with phantom passengers.

Although manipulating commission structures isn't unique to the airline industry, the early abuses were enough to derail Salter's efforts. The commission program at Continental ultimately came to a close and never caught on in the industry as a whole. Salter was right in recognizing that change was needed. Airlines couldn't continue to treat their most valued asset—their seat inventory—as an afterthought, but a different approach was required.

Like Salter, Bill Brunger observed the transition from inventory control to revenue management, but from a different perspective. A graduate of Middlebury College with a degree in liberal arts, Brunger spent three years teaching in the Brookline public school system near Boston before attending the Wharton School where he received an MBA. Degree in hand, Brunger spent the next six years at American, first as a senior pricing analyst and later as a manager of strategic planning.

At American, Brunger was involved with setting fares, not the operational pricing of opening and closing fare classes. Setting fares at American was handled much as it was everywhere else in the industry. Thick stacks of tariff sheets containing information from many different carriers would arrive at Brunger's desk on a daily basis. It was Brunger's job to review the fares in the markets he was responsible for and decide which fares American should change. Apart from the visually arduous task of searching through reams of paper, there was the obvious question of what circumstances should actually trigger a change. Said Brunger,

> Even a carrier as relatively sophisticated as American had made no preparation for the deregulation of fares. In 1980 and then more particularly in 1982 when full deregulation happened, we sat saying to ourselves, 'What the hell should we do?' Absent data and applicable theory, we floundered about thinking of heuristics that might help us decide which fares should go up or down based on what we knew about load factors and competitors. In the end, we matched downward moves in individual markets, and then occasionally took across-the-board increases to try to prop things back up.

Ludwig held a similar job for two years in the New York office of TWA, where there were two pricing analysts and two people dealing with the mechanics of filing fares. "It was the most boring job I've held in my entire life," said Ludwig. Even though Ludwig wasn't one of the analysts, he was extremely aware of the fundamental problem with setting fares. Who knew what the fares should really be? If a competitor dropped a fare from $300 to $280 for M-class travel from Saint Louis to Seattle, how should TWA respond? Should it match the $280 fare or leave its fare unchanged? The problem was further complicated by the fact that TWA might not have the equivalent of the competitor's M class. So if not M class, what class or classes should have their fares changed?

To this day, airline pricing departments grapple with the problem of setting fares. Though it often comes as a surprise that pricing (fare setting) is far less developed as a science than operational pricing (setting inventory availability), a closer look at the two problems reveals why this is the case. With a set of fixed fares for a given flight—say, $200 for Y class, $150 for M class, and $100 for Q class—operational pricing doesn't focus on whether Q class should be $100.00, $100.01, or some other number. It focuses on whether or not a Q-class ticket should be sold at all, or if it should be pulled from the shelf so that customers must pay $150—a considerably more tractable problem in practice.

While airlines happened upon this distinction because fare classes were already in place, fare classes with their many restrictions aren't necessary to perform the operational pricing practiced by airlines. If they chose, airlines could ignore restrictions and simply offer three different price points—$200, $150, and $100—then pick which of these prices they want to offer at any given time. This is the approach used by many of the low cost carriers.

What's important to realize, and is essential for effective pricing, is that the focus is on "one step up" or "two steps up," not on "up by $42.63," "up by $42.64," or some other number. Theoretically, operational pricing can benefit from finer and finer price granularity. Practically, it's easy to get lost in the search for the perfect number, a number that, even if it existed, no one would agree upon.

This isn't to say that setting fares isn't important, and airlines spend considerable time and effort on just this problem. Ben Baldanza, the president and COO of Spirit Airlines who has served in many airline executive positions, including as senior vice president of marketing and planning at U.S. Airways and executive vice president of marketing at Continental, likened airline pricing departments to a gas: they tend to fill whatever container they're placed into. The ability of pricing departments to expand shows just how seriously fare setting is treated by management. Pricing analysts have at their disposal the latest tools for reviewing, analyzing, and changing fares in the market. The net result of their efforts is strategic, not tactical: they set the rules of engagement, then let operational pricing groups play the game. Herein lies one of the hidden secrets of scientific pricing in the airline industry, one that even many people who work in the industry are not consciously aware of.

Pricing departments promote consistent fares relative to competitors. If a low-cost competitor enters a market and offers prices that are half the fares offered by a carrier already flying in that market, pricing departments are quick to respond by lowering overall fares. They may not hit the perfect number, but they help keep the playing field level, which in turn benefits the revenue management department, with its focus on inventory classes. Revenue management analysts can safely focus on whether "lower-priced" or "higher-priced" inventory classes should be available without getting overly concerned about the actual prices. And revenue management science can remain less complicated than it would otherwise be. While seemingly crude, the strategic/tactical interplay between pricing and revenue management departments has proven to be a good engineering approximation to a remarkably complex problem. And to a large extent, it's been the result of nothing more than happenstance, resulting from the fact that airlines sell discrete products called fare classes.

Although Brunger was never involved with revenue management during his time at American, he had the opportunity to observe the first true revenue management system in action, DINAMO—the Dynamic Inventory Allocation and

Maintenance Optimizer. DINAMO wasn't a reservations system, but sat along-side the reservations system. It used historical data gathered by the reservations system to predict how many passengers would arrive in the future for each flight in each fare class. After using this information to determine how many seats to make available in each fare class and how much to overbook the flight, it fed this information back to the reservations system. If the reservations system was a high-performance box for managing and keeping records of passenger bookings, DINAMO was the scientific brain behind the reservations system that told it who to accept.

DINAMO provided Crandall and his team previously unheard-of operational pricing power. American could offer inexpensive fare products to match or undercut its competition on a flight-by-flight basis, then limit the availability of these products if DINAMO concluded there were enough high-paying passengers to fill the seats. American could have its low-fare cake and eat it, too.

The approach proved to be a nightmare for American's competition. People Express, one of the most successful upstarts born in the era following deregulation, was devastated by American's efforts. CEO Donald Burr's admission is now legendary:

> We were a vibrant, profitable company from 1981 to 1985, and then we tipped right over into losing $50 million a month. We were still the same company. What changed was American's ability to do widespread Yield Management in every one of our markets. We had been profitable from the day we started until the day American came at us with Ultimate Super Savers. That was the end of our run because they were able to underprice us at will, and surreptitiously.

In 1986, Brunger took a position as director of domestic pricing at Continental, but before leaving American, he made the rounds, interviewing with many of the major airlines. "American was way ahead of everyone," said Brunger, echoing a sentiment felt throughout much of the industry at that time.

The situation at Continental was indicative of the rest of the industry when he arrived there. Brunger was once again in pricing, but, like Ludwig, found that the job just wasn't that interesting. Among other things, his manager wanted to review fares on flights that were expected to go out less than 40 percent full. Red pen in hand, Brunger would spend hours searching for flights that met this criterion. The rationale behind the 40 percent figure was that "it seemed about right."

Two years later, Brunger was ready for a change. In addition to pricing, his manager also had responsibility for a small revenue management group that had recently been formed. Brunger talked to the head of the revenue management group, and the two swapped jobs. Brunger now found himself as the senior director of revenue management, leading a total of five people.

The new position allowed Brunger's creativity to blossom. Revenue management didn't receive as much management attention as pricing, and he was free to work on the problem with little interruption. For two weeks, Brunger drew curves showing the rates at which different fare classes sold during different periods of the booking cycle, trying to determine the best way to predict demand.

At the end of the two weeks, Brunger found himself on a late evening flight to Hawaii to visit corporate clients. The plane was full, and Brunger was seated in a smoky middle seat near the back. In keeping with airline travel policy, he was wearing a suit and tie, in contrast with the many vacationers in shorts and floral prints. With the lights out, Brunger worked away under the glow of the small, green, overhead light, busily scribbling Greek symbols. Enough of his thoughts had congealed so that it was time to write a computer program to automate his work. The computer language he had chosen was APL, a short-lived product from IBM that was a favorite of the mathematically inclined because APL used Greek characters, it required a special computer keyboard. Arriving back in Houston, Brunger typed his program onto the computer. Continental had its first revenue management system.

For the next two years, Brunger would make changes and additions to his code, incorporating new science and business rules. Finally, in the summer of 1990, he realized his code had grown to a point where it was suffering too many problems, from speed of operation to inevitable bugs. Brunger located a hidden office away from corporate headquarters and set about to fix the program once and for all. Even CEO Frank Lorenzo, who, like many CEOs, wasn't shy about dominating the time of those who worked for him, didn't know where Brunger was.

After three months, Brunger emerged from his hideaway, successful in having reached a firm decision: he was going to buy a new revenue management system. The details that went into building and maintaining such a system weren't worth the effort. Brunger admits that there were aspects of the decision that were disappointing. Purchasing a system meant he could no longer have an idea and try it out the same day. Like all creative people, Brunger likes thinking and doing. Still, the advantages of working with a company that developed a fully functional system complete with training, maintenance, and upgrades outweighed competing objectives.

Brunger was not alone in his build-versus-buy quandary. Especially when it comes to pricing, companies tend to view their way of doing business as unique. Without question, every company has certain unique aspects to the way it does business, but an objective look across different companies within an industry shows more commonalities than differences. Further, systems purchased from software vendors embody the industry's best practices that come from the experience of working with many different companies. And good software vendors are always looking to improve their products in an effort to remain ahead of their competition. In rapidly changing fields such as scientific pricing, it's almost impossible for individual companies to keep up with advances unless they devote extensive internal resources to the task. Building and maintaining an internal system can prove to be extremely expensive and, as United and TWA discovered with their first reservations systems, extremely risky.

Brunger soon moved on to hold many more senior positions in pricing and revenue management, but the build-versus-buy decision is aptly captured by the two bullet points on his resume describing his accomplishments during his three years as senior director of revenue management.

- Designed and created all programming code for Continental's original revenue management system
- Subsequently negotiated outsourcing of revenue management technology

* * *

The floor is much as you'd find in any high-rise office building in a major city. Paneled elevators open onto a central corridor with heavy glass doors at either end. The doors are locked and require a keycard or a friendly employee to gain entrance. The open floor plan, filled almost entirely with cubicles, is cleaner and more inviting than most, perhaps because the blue, gray, and white motif is both businesslike and calming at the same time. The only truly unique feature of the building is that the east wall looks directly upon the former Enron headquarters, a constant reminder that to survive, a company needs to remain focused on the bottom line.

The floor and the floor below it are home to the revenue management, pricing, and scheduling departments of Continental Airlines. Among these groups, revenue management is the largest, comprised of some 85 people. As an established department, its responsibilities have grown over the years, and not all of the individuals who work there concern themselves with opening and closing fare classes. Some keep an eye on valuable BusinessFirst seats, confirming reservations and making sure no seat goes out empty if there's someone who wants it. Others track down instances in which savvy travel agents have made bookings that don't adhere to company policy. Still, the focus of over half the group is on revenue management—determining how much to overbook flights and how many seats to make available in each fare class.

Analysts are divided into teams based on the geography of the flights they attend to. Less experienced analysts work closely with their managers as they learn how to manage special situations, while experienced analysts have a much higher level of autonomy and simply report to their managers what actions they've taken. Strategic analysts may not deal directly with inventory levels at all, instead focusing on higher-level market issues.

The number of seats in each fare class is determined by a revenue management system, and all but a few percent are fed directly to the reservations system without being reviewed by an analyst. However, in business, there's no such thing as a completely normal day, and the system flags those flights that are not behaving as expected for an analyst to take a look at.

Manny was just concluding an investigation when I sat with him. Reviewing a weekly report, he discovered that for flights over the coming months, bookings into a particular city were abnormally low compared to the prior year. As a starting point for his analysis, he had information at a very aggregate level—all flights into and out of the city at a weekly level. From there, he was able to dig down further until he ultimately located the problem. The number of bookings from a particular foreign city had dropped precipitously. Manny explained that in the prior year, there had been a code-share agreement in place with a foreign carrier. The foreign carrier booked passengers to the city Manny was investigating, but flew them only halfway before they changed to a Continental flight. With the code-share agreement no longer in place, this source of traffic had all but disappeared. Had Manny been unable to explain the situation as a result of his efforts, he would have communicated the problem to a market analyst. The market analyst would then have looked in more detail at how the inventory levels were actually being set, and, if necessary, modify parameters affecting system calculations.

Manny's example is typical of how revenue management systems and analysts rely on one another. Forecasting uses historical data to predict future passenger demand, but when something out of the ordinary occurs, history becomes an even less perfect indicator than it was to begin with. Good revenue management systems support analyst intervention with facilities for addressing holidays, promotions, and other special events. The emphasis is on providing a means to help the revenue management system understand what's going on. Analysts aren't encouraged to bypass the revenue management system and pick the inventory levels that go into the reservations system. Instead, they analyze special cases in which bookings aren't proceeding as they should, determine if corrective action is required, and, if it is, use data and good business judgment to take action.

Analysts are important to the operation of any pricing scheme, yet the sirens of science can obscure the obvious. Science conjures up notions of infallibility; that there's a right price and science will find it. Organizations that enter into a pricing project with such high expectations are doomed from the outset since they can never be met. In the same way that weather can't be predicted with certainty, pricing science faces its own set of limitations. Yet just as we successfully use weather predictions to plan our affairs, pricing systems can be powerful tools. Making these tools work requires properly integrating analysts into the pricing process.

For their part, revenue management systems handle most cases far better than their human counterparts. With access to data and a coolly rational approach to decision-making, they are neither overly optimistic nor pessimistic. They can employ reasoning in the form of scientific methods that humans aren't capable of without computational assistance. Revenue management systems also free analysts from the routine activities that historically dominated their time.

The role of today's revenue management analysts is light-years beyond the guesswork of early inventory controllers. Recognizing this, Continental invests in education. New analysts are not only taught about the airline business and the systems they'll be required to work with, but they also receive a refresher course in statistics and an introduction to the mathematical methods used to calculate how many seats are assigned to each fare class. With an emphasis on working with the revenue management system rather than fighting it, analysts are taught to understand and address the root concern rather than to slip into the bad habit of overriding inventory levels they don't agree with. If something occurs that impacts the quality of the forecasts, it's the forecasts that need attention. Roughly half the analysts hold MBAs. Not all airlines maintain such high standards, nor is it necessary in all cases. But the hiring and training practices used by Continental reflect the importance the company places on revenue management.

In observing the day-to-day activities of revenue management analysts, rarely does price enter into their discussions. They are, of course, acutely focused on filling aircraft and doing so with as many people in high-priced fare classes as possible. But price isn't their primary concern. Like the relationship between chips and money in a Las Vegas casino, fare classes have value, but the associated price only lingers in the background as analysts work with them. Next week's flight should expect to have at least 20 full-fare customers because it had that number last week, the week before, and the week before that. To send a flight out with fewer would be playing the game incorrectly. Good revenue management analysts want to win the game.

* * *

Continental and airlines like it represent the pinnacle of an evolutionary chain. The need to keep track of how many seats were left on a plane led to ever more sophisticated means of keeping records. The need was so great that it forced airlines to push the limits of computer technology. As they became proficient at internally managing reservations, they grew increasingly aware of the power of their newly built systems to sell tickets. Driven by intense competitive pressure to capture and own market share, airlines with reservations systems were putting computer terminals on the desks of thousands of travel agents. This was as close as any industry had come to providing a direct electronic connection to a vast number of consumers, and with their power to pick and choose which flights were offered to passengers, for all practical purposes, travel agents were acting as consumers. While the Internet would provide the final producer-to-consumer link, by setting up their early electronic distribution network, airlines would deal with e-commerce decades before the term came into use.

For its part, revenue management evolved from a realization that inventory controllers had a big impact on revenue. But the science of revenue management could only develop once reservations systems were in place to systematically collect historical data in a form amenable to analysis. Yet as the many players discovered, data and a recognized business problem were only the starting point. Conviction and commitment were necessary to elevate inventory control—operational pricing—to a role in which it could promote the financial health of airlines.

Notes

- More information on the history of the airline industry's pioneering work with computers and reservations systems in particular can be found in the following sources, all of which were used in the preparation of this chapter: Copeland and McKenney (1988), Copeland, Mason, and McKenney (1995), and Eklund (1994). Petzinger (1995) provides a more general account of the overall history of the airlines, including reservations systems.
- The quotation by Donald Burr is taken from Cross (1997, p. 125).
- In addition to the individuals named in the text, Lee Davis, Leon Kinloch, Greg Lough, and Manny Sousa all provided source material contained in this chapter.

CHAPTER 4

How It All Works

The late 1980s were an exceptionally good period for director Oliver Stone. Winning Academy Awards as best director for the films *Platoon* (1986) and *Born on the Fourth of July* (1989), Stone won accolades but was ultimately passed over for the equally provocative film *Wall Street* (1987) in which he analyzes the inner workings of fictional corporate raider Gordon Gekko. Michael Douglas, who won the Academy Award for best actor portraying Gekko, captures the no-holds-barred capitalistic spirit of the film with his shocking but memorable line, "Greed, for lack of a better word, is good."

Less well remembered is the exchange between Gekko and the union heads at Bluestar Airlines. Seeking union support as he attempts a hostile takeover bid, Gekko turns to his protégé, Buddy, played by Charlie Sheen, to present a plan for saving the airline.

Gekko: You've got losses of $20 to $30 million, dividends cut to zero, and you're being squeezed to death by the majors . . .

Head of flight attendants union: What's your marketing strategy? How do you intend to return us to profitability?

Gekko: Why don't I give Buddy an opportunity to answer that. Buddy?

Buddy: . . . Now, what I've come up with here is a basic three point plan, alright? One, we modernize. Our computer software is dog shit. We update it. We squeeze every dollar out of each seat and mile flown. You don't sell a seat to a guy for 79 bucks when he's willing to pay 379. Effective inventory management through computerization will increase our load factor by 5 to 20 percent. That translates to approximately 50 to 200 million dollars in revenue. The point being, we can beat the majors in a price war.

By the late 1980s, airline pricing was clearly in the focus of the public eye. Whereas a decade earlier, prices were stable and uniform, the new freedom provided by deregulation caused a proliferation of ever-changing fares. One of the big, if inevitable, misunderstandings in the market was that the many different fare products were introduced to force business travelers to pay higher prices. Quite the opposite was true. Under regulation, prices were sufficiently high so that planes routinely flew with far more empty seats than they do today. With an inability to compete on price and with planes half empty, the thought of in-flight piano bars to attract more customers actually made sense. As airlines exited the regulatory era, their first thought wasn't to raise prices, but rather to fill the empty seats without dropping the price for business travelers. This is why Buddy places emphasis on revenue improvements coming from increased load factors—the ratio of the number of passengers on a flight to the actual number of seats.

It was the drive to fill empty seats that led to ever more elaborate ways of managing seat inventory. Defining different fare classes led to a realization that controlling fare-class inventory could have huge financial repercussions, which in turn led airlines to ask, "What, *exactly*, is the most profitable way to control inventory?" The basic answer is simple enough. If a flight has one seat left a week before it departs, the airline can expect to get a high price for it. If the flight has 100 seats left a week before it departs, it will probably need to offer a much lower price if it wants to fill the plane. The more difficult answer lies in finding the actual number of seats to leave open in each fare class.

While revenue management is serious business, it can also be thought of as a game. In my bag I hold an unknown number of marbles, some labeled with Y and others labeled with Q. The marbles labeled Y are worth $500 and the marbles labeled Q are worth $200. I'm going to take marbles from my bag and hand them to you one at a time, showing you the Y or the Q. When I hand you a marble, you have to make a decision then and there if you want to keep it. If you reject it, we throw it away. If you keep it, I'll pay you the amount stamped on that marble at the end of the game. You get to keep a total of 100 marbles. What decision process do you use to decide what to keep and what to reject?

Before playing, you may want to ask me some questions. For example, how many Y and Q marbles are in the bag? If I told you, the game would be easy. With the information that "there are 75marbles marked with a Y," you'd take any Y marble and no more than 25 Q marbles. In this way, when the game was over, you'd have all of the Y marbles and as many Q marbles as possible. (Notice that you didn't actually have to know how many Q marbles were in the bag.)

If I wouldn't tell you exactly how many Y and Q marbles were in the bag, you might ask if I'm going to pull the marbles out in any particular order. If by chance I told you I was going to pull all the Y marbles out before all the Q marbles, you'd be in great shape. Accept every marble until you reach the limit of 100, and again you'd be sure to make as much money as possible.

Even if I wasn't going to give you all the Y marbles first, I might nonetheless give you useful information. If I told you I would pull all the Q marbles out before any of the Y marbles, it would be a very different problem than if I told you I would pull the marbles out of the bag at random. In both of these cases, the best decision process isn't obvious, and without knowing exactly how many Y and Q marbles are in the bag, there isn't a decision process that always guarantees the most possible money. But in each case there is a decision processes that leads to the most money on average, and the process is different depending on the order in which the marbles will be pulled from the bag.

Replacing "marbles" with "passengers," this is exactly the game faced by airlines. The game becomes even more complicated when there are more fare classes, passengers can cancel, and when a plane has more than one cabin (for example, a first-class cabin in addition to coach), but the fundamental concepts can be explained with two fare classes, no cancellations, and a single coach cabin. Airlines ask questions about marbles in the bag using historical data to forecast who they expect to show up and in what order. Similar flights in the past hold valuable information about what to expect in the future. Airlines don't look at each passenger and say "accept" or "reject," but rather do something similar, using inventory limits kept on their reservations systems to manage who can and

---

Sidebar 4.1: The Secretary Problem

An interesting variation of the marble game is known as the *secretary problem*. In this game, each marble has a different number attached to it. You are given no information whatsoever about the numbers on the marbles, but you are told that there are n of them in the bag. Your goal is to find the marble with the largest number attached to it by looking at the marbles as they are randomly drawn from the bag. As each marble is shown to you, you must either declare that it's the marble you think is the largest, or you must discard it and keep drawing. The name *secretary problem* comes from interpreting the marbles as interview candidates for a secretarial position, and the attached number as the quality of the candidate.

One strategy is to reject the first k marbles, where k is a number to be chosen, then declaring the first marble with a bigger number than you've seen on all of the first k marbles as your selection. For example, if there are $n = 100$ marbles, you might decide to reject the first $k = 10$, finding that the largest number you observed was 7,945,272. Continuing to draw, you find that the 47th marble is the first one you encounter with a number bigger than 7,945,272, and you declare it to be the largest number in the entire bag of 100 marbles. Remarkably, if k is chosen as approximately $0.36 \times n$, this strategy yields the marble with the largest number attached to it over 36 percent of the time, even if there are millions of marbles (or more) in the bag. Put another way, since you have better than a 1/3 chance of finding the largest number, if someone is willing to bet $2 to your $1 that you can't find it, you're making a good bet.

can't get a seat. With two fare classes and 100 seats, an airline might set aside 30 for Y class and 70 for Q class. If a passenger shows up and wants to buy a Q-class ticket, he's free to purchase it as long as there's inventory available. If he does make a purchase, the inventory is reduced by one.

Of course, with inventory partitioned in this way and only a forecast of who will show up, an airline may find itself in the unfortunate position of having sold out the high-priced Y-class product while still having the low-priced Q-class product available. If this occurs, someone wishing to buy a Y-class ticket would be turned away even though there are still seats left on the plane.

To avoid this, airlines typically use *nested* inventory levels. With nested inventory levels, any seats available for lower priced products are also available for higher priced products. The nested inventory levels shown in Table 4.1 amount to saying, "Accept all Y-class passengers as long as you have an empty seat, but accept no more than 70 Q-class passengers." The nested inventory levels are also shown after the arrival of one Y-class passenger and one Q-class passenger. Nested inventory levels define a strategy for accepting and rejecting passengers, but not all strategies can be implemented using nested inventory levels. We return to discuss the limitations of reservations systems and nested inventory levels in the next chapter.

To determine how to set the nested fare class levels, airlines forecast how many passengers they expect will come looking for tickets in each fare class. If on average 30 people showed up looking for Y class tickets in the past, a good basic estimate of the number of people who will show up next time is 30.

So does this mean that an airline should protect 30 seats for Y-class passengers on its one 100–seat plane, using a nested inventory level of 100 for Y class and limiting Q-class purchases to 70? Without any further information, this approach is reasonable. However, airlines don't just forecast the average number of Y-class passengers they expect to arrive. Instead, they forecast the probability that 10, 20, or any number of passengers will show up. For example, in Table 4.2, we see that there is a 51 percent chance of 29 or more Y-class passengers appearing, while there is only a 44 percent chance of 33 or more passengers appearing.

Table 4.1 An example of nested inventory levels for a 100-seat aircraft. After the arrival of a Y-class passenger, the Y-class inventory level is reduced. After the arrival of a Q-class passenger, both the Q-class inventory level and the inventory level of all higher-priced fare classes are reduced, reflecting the fact that there is one less seat on the aircraft.

Fare class	Price	Available seats	Available seats following a Y-class arrival	Available seats following a Y- and a Q-class arrival
Y	$500	100	99	98
Q	$200	70	70	69

Table 4.2 A list of the number of Y-class passengers that might arrive together with the probability that at least that many passengers actually do arrive. The figures in the row labeled *expected marginal revenue* are calculated by multiplying the Y-class fare of $500 by the associated probability. The expected marginal revenue for Y class is at least as large as the Q-class price of $200 up to 34, leading to the nested inventory levels shown.

Number of Y-class passengers	...	29	30	31	32	33	34	35	36	...
Chance of at least this many Y-class passengers	...	51%	50%	49%	47%	44%	40%	35%	30%	...
Expected marginal revenue	...	$255	$250	$245	$235	$220	$200	$175	$150	...

Fare class	Price	Available seats
Y	$500	100
Q	$200	66 (100 − 34)

Information about the likelihood of how many people show up proves to be extremely useful in determining how many seats to protect for Y-class tickets. With a ticket selling for $500, if 30 Y-class tickets are set aside, how much do we expect to get for the thirtieth ticket? If we sell it, we get $500, but there's only a 50 percent chance that 30 or more people will show up and that it's sold. It seems reasonable to weight the price of the ticket by the chance of selling it, and in this case, $500 × 50 percent = $250. The value $250 calculated in this way is known as the *expected marginal revenue* of the thirtieth seat.

Expected marginal revenue, often shortened to just *marginal revenue*, is the single most important concept in all of airline ticket pricing. It's such a fundamental concept that it carries a variety of names depending on the industry and the context in which it's being used: opportunity cost, displacement cost, bid price, and hurdle rate are just a few. The term marginal revenue derives from adding a seat to a class, as in the example, "If I have 29 seats set aside for Y class, what's the marginal revenue I can expect to make from adding a thirtieth?" Opportunity cost is just the opposite. "What is the opportunity cost of taking away the thirtieth seat from Y class and moving it elsewhere?" The same holds true for displacement cost. Bid price and hurdle rate are used in the context of selling a seat. "I'll take $300 for that seat because it is more than the bid price of $250." It's not uncommon to hear someone use two or three of the terms in the course of a discussion as the exact meaning subtly changes.

How do marginal revenues help us calculate how many seats to save for Y-class passengers? Notice from our calculation of marginal revenues that the thirty-fourth seat is worth $200, exactly the same amount as our Q-class seat. Beyond that, each seat set aside for Y class is worth less than a Q-class seat. It therefore seems reasonable to set aside 34 seats for Y-class passengers.

This simple method is known as *Littlewood's rule*. It was the first scientific approach to determining fare-class availability, and variations on the theme still serve as the foundation for many different revenue management systems. The method makes intuitive sense, and years of experience have shown that it works well.

* * *

"Optimal" is a frequently misunderstood word. In our everyday lives, we use optimal to describe something that's the best. We buy car A instead of car B because it's the optimal choice. We take the bus rather than walk because it's the optimal way to get to work. Buried in our notion of optimal are unstated objectives, constraints, and assumptions. Our objective may be getting to work quickly, with constraints on how much we're willing to spend. We then need to make assumptions about how dependable the bus is. Does it arrive every 30 minutes like clockwork, or is the service unpredictable?

To management scientists, optimal has a very specific meaning: the best solution *relative to a stated set of objectives, constraints, and assumptions*. In this sense, a scientist would concede that in making a decision to ride the bus, we've weighed the various options and come to a good, maybe even very good, decision. But we haven't optimized anything since we haven't clearly articulated our objectives, constraints, and assumptions—a requirement to make any formal statements about optimality.

Littlewood's rule is optimal for determining nested fare class levels, but only under the assumption that all of the Q-class passengers arrive before any of the Y-class passengers start showing up. This may appear to be a rash assumption, since passengers certainly don't show up in this nice, orderly fashion. However, in many cases, this Q before Y assumption is a reasonable approximation, and as a result, the method works reasonably well. Of course, more advanced methods exist for different assumptions about how passengers arrive, but from a historical, practical, and conceptual standpoint, Littlewood's rule was a milestone.

Taken together, the objectives, constraints, and assumptions we settle on when trying to solve a problem comprise a *model* of the problem. Creating a model is important because it forces us to clarify what we're facing and what we're trying to do. Another important aspect of creating a model is that it allows us to convert the problem into a form that makes it easy to find an optimal solution. An example helps make this clear.

The CFO Problem

The CFO of a division of a large corporation is evaluating five different investments. If undertaken, each investment requires a capital outlay now and an additional capital outlay in six months, with revenue generated twelve months in the future. For example, project one requires a capital investment of $11 million now and $3 million six months from now, and will generate $16 million in twelve months.

The investments along with their capital outlays are shown in Table 4.3. The amounts of capital the CFO has available to spend now ($26 million) and in six months ($12 million) are also shown in this table. Capital that is not used now may not be carried forward and invested in six months, since these amounts will be allocated to other projects not under the purview of the CFO.

A project may be undertaken in a fractional amount at a fractional cost, but it also generates a fractional return. For example, the CFO may choose to take half a position in project 1 at a cost of $5.5 million today and $1.5 million in six months, but it will only realize a return of $8 million. Of the five potential projects, what positions should the CFO take in order to maximize the revenue from these projects one year hence?

Not only do we have a problem, but we also have a model because there are clearly described objectives, constraints, and assumptions. There is an optimal solution, though we may not immediately see what it is. How do we find it?

We might begin by ordering the projects by the amount of revenue they generate and keep undertaking projects until the available capital is consumed. Project 5 has the highest value, but note that it requires the entire capital budget of $12 million available in six months. Thus, if project 5 is undertaken in its entirety, it is the only project that can be undertaken, and the revenue generated in twelve months will be $40 million.

Since the capital available for investment in six months is limited, an alternative approach would be to order the projects from least to most capital used in

Table 4.3 Amounts in millions of dollars

	1	2	3	4	5	Capital available to invest
Revenue project will generate after 12 months	16	16	15	14	40	—
Investment required now	11	10	5	5	10	26
Investment required in 6 months	3	4	5	3	12	12

this period, and to keep undertaking projects until the available capital is consumed. Projects 1 and 4 in either order use the least capital after six months, followed by projects 2, 3, and 5. Note that projects 1, 4, and 2 can be undertaken in their entirety, but together they use all $26 million of the capital available now. However, the combined revenue of these projects is $46 million—an increase of $6 million over the previous solution.

Our ad hoc approach seems to be working, but there must be a better way. Even if we do happen upon an optimal solution, how do we know it's optimal? Management scientists deal with this difficulty by first converting the model into mathematics. To do so, we begin by introducing a variable for each project. For example, x_1 might represent how much of project 1 to undertake. If x_1 is 0, we don't undertake the project, if x_1 is 1, we undertake the project in its entirety, and if x_1 takes on a fractional value, we undertake that fraction of the project. With these variables, it's now possible to create a mathematical model as shown in Table 4.4.

The expression to the right of the word "Maximize" is the *objective function*, and it gives the revenue associated with any solution. The remaining expressions represent the constraints. If we choose values for the variables that satisfy these constraints, then the corresponding choice of projects can be undertaken. Likewise, every possible project can be represented by an appropriate choice of variables. The best solution we've found thus far is represented as $x_1 = 1$, $x_2 = 1$, $x_3 = 0$, $x_4 = 1$, and $x_5 = 0$.

One benefit of creating a mathematical model is that it can be given to a computer. The computer understands the mathematical objective and constraints and can be told to look for a solution. It doesn't care where the objective and constraints came from. Knowing that the model is in this mathematical form, it will look for an optimal solution. Understanding the methods by which computers find optimal solutions requires a background in mathematics, and management scientists devote considerable energy to developing ever better methods.

Giving the CFO's problem to a computer, we find that the optimal solution is $x_1 = 1$, $x_2 = 3/4$, $x_3 = 0$, $x_4 = 1$, and $x_5 = 1/4$, with an associated revenue of $52 million. Many important lessons can be learned from this example.

First, it's not immediately clear why this is the optimal solution. We can verify that the solution satisfies the constraints and that it has a value of $52 million, but why undertake 3/4 of project 2 and 1/4 of project 5? Unfortunately, while we can usually convince ourselves that the solution at least seems reasonable, there

Table 4.4 CFO problem expressed as a mathematical model

Maximize	$16 x_1$	$+ \ 16 x_2$	$+ \ 15 x_3$	$+ \ 14 x_4$	$+ \ 40 x_5$	
Subject to	$11 x_1$	$+ \ 10 x_2$	$+ \ 5 x_3$	$+ \ 5 x_4$	$+ \ 10 x_5$	$\leq \ 26$
	$3 x_1$	$+ \ 4 x_2$	$+ \ 5 x_3$	$+ \ 3 x_4$	$+ \ 12 x_5$	$\leq \ 12$
	$0 \leq x_1 \leq 1$	$0 \leq x_2 \leq 1$	$0 \leq x_3 \leq 1$	$0 \leq x_4 \leq 1$	$0 \leq x_5 \leq 1$	

isn't always a good way to see when a solution is optimal just by looking at it, nor is there always a good common sense way of finding it.

It's also important to realize that it's not necessary for someone to do the work to actually find an optimal solution. If the model and mathematical model properly represent the underlying problem, the computer will find it. While it can be disconcerting to relinquish the process of finding an answer to a computer, failing to do so can be extremely expensive. A $6 million increase over a $46 million solution represents an improvement of 13 percent. Many of us quickly skip over carefully understanding the constraints and assumptions in our decision making and jump right to solving an ill-defined problem. Management scientists do just the opposite. Faced with making a decision, they spend all their time on the constraints and assumptions and let the computer do what it's good at—solving the problem. As problems get bigger, it becomes impossible for anyone to account for the many conflicting constraints and assumptions. Today's computers could easily handle a problem for our CFO with tens of millions of projects and millions of constraints.

* * *

Although the conceptual foundation of communism stems from the philosophy of Karl Marx, Vladimir Lenin was responsible for bringing it to life. Marx voiced a philosophy; Lenin had to figure out how to convert it into a functioning form of government. Yet on the eve of the October Revolution, it was clear that Lenin had yet to work out the details. In a letter written only hours before Bolshevik troops moved on the city of Petrograd, he penned, "The point of the uprising is the seizure of power; afterwards we will see what we can do with it."

Soviet mathematician and economist Leonid Kantorovich was one of those individuals tasked with figuring out what to do with the Soviet economy. "For the first time in history," said Kantorovich in his acceptance lecture for the 1975 Nobel Prize in Economics, "all the means of production passed into the possession of the people and there arose the need for the centralized and unified control of the economy of the vast country." Unlike capitalist economies that use prices established by the free market to determine what gets produced (high prices inducing greater production, low prices, less), production in the Soviet economy depended on direction from the Communist Party. Someone, somewhere, had to decide what got made and who got to use it.

For Kantorovich, central planning presented a new way of looking at economic problems as he drew upon his formal education as a mathematician. In the West, mathematics was rapidly gaining prevalence in the study of economics. If economic relationships could be expressed in mathematical terms, then economic questions could be answered through clearly articulated deductive reasoning. Economists could agree to disagree about their mathematical assumptions, but there could be no argument about where the assumptions would ultimately lead.

Although the problems Kantorovich and his colleagues faced were orders of magnitude larger than the CFO problem, the basic idea was the same. The Soviets had limited resources in the way of factory capacity, natural resources, and labor, and the Communist Party had objectives it wanted to achieve in the way of food and material production. The question was how to maximally achieve their objectives while satisfying their limited-resource constraints.

Kantorovich first developed a mathematical method for finding optimal solutions to problems like the CFO problem while working for the Laboratory of the Plywood Trust. In 1939, just as Stalin was completing the most intense half-decade of purges during his reign over the Communist Party, Kantorovich's method was published in a booklet entitled *The Mathematical Method of Production Planning and Organization*. Given the prevailing climate in the Soviet Union, the work remained unknown to Western scholars for many years.

In discovering his method, Kantorovich also discovered an extraordinary property of problems like the CFO problem. When searching for an optimal solution, his method also found special *shadow prices*, one for each resource constraint. In the CFO problem, there are two such constraints, the $26 million of capital available now, and the $12 million of capital available in six months. The shadow prices for these constraints are $0.4 million and $3 million, respectively.

What are these shadow prices? To answer this question, try using them to calculate the profitability of each project the CFO is considering. For example, using the shadow prices, what is the cost of undertaking the first project? The project uses 11 units of capital now and three units of capital in six months. The "cost" of this project using the shadow prices $(11 \times 0.4) + (3 \times 3) = \$13.4$ million. The return on the project is $16 million, so that the "profit" is $16 million $- \$13.4$ million $= \$2.6$ million. At least with respect to the shadow prices, we find that the first project is profitable.

Notice that in our optimal solution to the CFO's problem $(x_1 = 1, x_2 = 3/4, x_3 = 0, x_4 = 1,$ and $x_5 = 1/4)$, we've chosen to undertake the first project in its entirety. Relative to our shadow prices, this makes sense. If the project is profitable, we should undertake as much of it as possible. Expanding on this basic idea of measuring profitability, it turns out that if we have a proposed solution for the CFO's problem and can find shadow prices meeting the following profitability conditions, then our proposed solution *must* be optimal (the fact that this is true is not obvious and requires mathematical derivation):

1. Projects that are being fully undertaken are profitable (or have a profit of 0).
2. Projects that are not being undertaken are unprofitable (or have a profit of 0).
3. Any projects undertaken in a fractional amount are break-even.

Looking at project 2, we find that the optimal solution has $x_2 = 3/4$. The cost is $(10 \times 0.4) + (4 \times 3) = \$16$ million, and the profit is $16 million $- \$16$ million $= \$0$, as expected. Project 3, which is not being undertaken, has the cost $(5 \times 0.4) + (5 \times 3) = \$17$ million, and the profit $15 million $- \$17$ million $= -\$2$

million. Projects 4 and 5 also meet the profitability conditions, and so the proposed solution to the CFO's problem is optimal.

Not only are the profitability conditions sufficient for showing that a proposed solution is optimal, but it turns out that they are necessary as well. If you have a proposed solution, and you can't find a set of shadow prices meeting the conditions, the proposed solution can't be optimal. Some of the most successful methods for finding an optimal solution work by first proposing a solution and a set of shadow prices, then checking the profitability conditions. If they are met, the proposed solution must be optimal. If not, then the failure can be used to construct a new and better solution by undertaking more of a profitable project or less of an unprofitable project. In the process, new shadow prices are generated, and the process continues.

For Kantorovich, the realization that solving central planning problems is intimately linked with establishing prices for the productive resources—that, in fact, there is no good way to avoid generating prices—posed a politically sensitive problem. Met guardedly by many Soviet economists, Kantorovich's earliest published results did not mention prices but instead borrowed from mathematical terminology and referred to "resolving multipliers." He would later use the terms "synthetic indices" and "objectively determined valuations." In each case, there was an effort to avoid terms that could be interpreted as anything related to the prices found in a free market.

Yet Kantorovich clearly understood the implications of his work. Throughout his life, he focused attention on how prices could be used to help decentralize decisions in a centrally planned economy. In essence, rather than telling people what to produce, he promoted telling them how much things cost and allowing them to make their own production decisions.

Fifteen years before the fall of the Berlin Wall, Kantorovich declared before the Nobel Prize Committee that a consistently planned economy cannot do without indices analogous to the prices, rents, and interest rates that are directly observed in capitalist economies, an admission that could have led to his deportation or execution in earlier years. The prize committee, too, was clear on why Kantorovich was receiving the award, honoring him for "[demonstrating] the connection between the allocation of resources and the price system."

To claim that Kantorovich was bestowed with the Nobel Prize as a way of prodding the Soviet Union toward a free market economy would be far too simplistic. Although Kantorovich's work was not initially greeted with open arms in his home country, over the years it gained ever stronger acceptance. In 1965, Kantorovich was honored with the highly prestigious Lenin Prize, an award bestowed for achievements in the arts and sciences every even year on Lenin's birthday. Kantorovich was also inducted into the Order of Lenin two years later, and went on to receive many other honors. When he received his Nobel Prize, Kantorovich was heading the research laboratory at the Institute of National Economy Control in Moscow, where he acted as consultant to various governmental agencies and high-ranking Soviet executives. While the Soviet Union was certainly not going to fully embrace free markets, it was coming to terms with the importance of prices as a fundamental concept, even in a centrally planned economy.

Kantorovich was far from alone in his research on optimization. In the United States, Tjalling Koopmans would later do work along the same lines as Kantorovich. Unaware at the time of the work being performed in the Soviet Union, Koopmans would later share the Nobel Prize with Kantorovich. In fact, Koopmans took the initiative to have a translation of Kantorovich's 1939 booklet published in the journal *Management Science*.

Kantorovich and Koopmans are only part of the much grander story of optimization, a story involving many of the most prominent names in twentieth-century mathematics and mathematical economics, including George Dantzig and John von Neumann. Though many of the fundamental ideas related to optimization had been known in different forms for centuries, the middle to late twentieth century proved to be the most active and exciting time ever experienced by the field, first as economists embraced mathematics, and then as computers opened new computational realms.

And wherever they turned, researchers kept bumping into shadow prices. Like the many different designations used by Kantorovich, these prices were given different names depending on where they popped up—simplex multipliers, Lagrange multipliers, dual prices—all referring to essentially the same thing. And if their role as prices wasn't enough, they also received a completely different set of names that arise from yet another remarkable interpretation.

Shadow prices also represent the *marginal value* of each constraint's resource relative to the optimization problem. The resource being used in the second constraint of the CFO problem is the capital available six months from now. The shadow price of $3 million means that if $1 million were available six months from now, increasing the available capital from $12 million to $13 million , the optimal solution would increase from $52 to $55 million. The marginal value of an additional $1 million six months from now is $3 million.

If the amount of capital decreased from $12 million to $11 million , the optimal solution would decrease from $52 million to $49 million. Relative to the stated problem, capital is dear in six months, and if the CFO could find some extra capital for that period, he would get a three to one return on it.

Alternatively, the shadow price on the first constraint, representing the amount of capital available now, is $0.4 million, meaning that a $1 million change up or down in the available capital would have only a corresponding $0.4 million effect on the value of the optimal solution. As a result, the CFO might want to consider reducing the available capital invested in the project now.

It's worth stepping back to consider just how amazing these shadow prices are. For any constraining resource in an optimization problem, the shadow price represents exactly what that resource is worth relative to the entire optimization problem. While this may not seem too exciting when looking at the

CFO problem, imagine being able to pinpoint the marginal value of one particular resource in a problem with thousands of projects and constraints. A flexible manufacturing facility can determine the marginal value of everything from raw materials to time at each station. A package delivery company can determine the marginal value of adding truck capacity or hiring more people. A hospital can determine the marginal value of adding inpatient rooms. And in each case, these marginal values are measured against the impact on the entire system. No longer is it necessary to guess what each part of a system is contributing. By clearly articulating the objectives, constraints, and assumptions of the underlying problem, a computer can not only provide the optimal course of action, but can also value each constraining resource.

Are optimization methods and their ability to generate solutions and shadow prices something to marvel at? Absolutely. Even with a problem as small as that faced by the CFO, the practical power of optimization becomes clear when an unintuitive optimal solution is found. The fact that we can now solve problems of almost unimaginable size borders on miraculous, and the underlying mathematical realm known as *duality theory* is filled with wonderful results and interpretations.

Yet it's the very power of optimization methods that's kept them from becoming more firmly rooted in business. Optimization methods look at every possible solution to a problem, and as a result, they frequently find optimal solutions that are not intuitive or, in the words of the practitioners who don't understand them, "don't make sense." The CFO problem is small enough that we can at least convince ourselves that the solution is probably right even if we don't understand it, but for larger problems, this task is often far more difficult.

When personalities are factored into the equation, even the results of small optimization problems can be called into question. Decision makers, not surprisingly, achieve their status by making decisions. An optimal solution that doesn't replicate what the decision maker would have done may be rejected out of hand or, more commonly, the model is challenged and discarded. The author has listened to some very creative arguments about why the CFO shouldn't use the optimal solution.

For airlines looking to manage prices by controlling fare-class availability, the marginal value of a seat is nothing more than the marginal revenue it generates. In finding their way onto the airline industry landscape, shadow prices developed the many new names associated with marginal revenue discussed earlier in this chapter. But their appearance wouldn't be fully appreciated until the airlines broke free of the direct city-to-city service, which dominated business under regulation, and began operating more complex flight networks.

Notes

- The Secretary Problem is known under a variety of different names. For a description and solution of the problem, see http://en.wikipedia.org/wiki/Secretary_problem.
- Littlewood's original paper describing his method for setting nested inventory levels was written in 1972 and presented at a meeting of the Airline Group of the International Federation of Operational Research Societies in that same year. A reprint of this award-winning work can be found in Littlewood (2005).
- The quotation by Lenin is from his letter to the central committee members, dated October 24, 1917. The translation is taken from Wolfe (1964).
- Details of the methods for calculating optimal solutions and shadow prices for problems such as the CFO problem are beyond the scope of this book. Interested readers are directed to any one of a number of introductory texts on the subject, such as Chvatal (1983), Dantzig (1998), Gass (1990), or Hillier and Lieberman (2004).

CHAPTER 5

When Passengers Collide

The rationale behind hub-and-spoke flight networks is simple enough. There may not be enough people wanting to fly from Spokane, Washington, to Fargo, North Dakota, to warrant a daily flight. However, there may be plenty of people who want to fly to Fargo from *somewhere* in the United States. If they're all gathered at one location and flown to Fargo together, the flight from this hub city to Fargo can be filled.

Of course, to make a hub-and-spoke system profitable, an airline needs to fill all of the flights into the hub as well. To achieve this, a network must be constructed with many different flights arriving and leaving the hub at roughly the same time. And the cities chosen for service must be such that they mesh well with the other cities in the network, with every flight in and out of the hub carrying enough passengers to make the network profitable.

While hub-and-spoke systems offer many advantages to both airlines and passengers, they greatly complicate the many aspects of running an airline, not the least of which is determining what revenue management methods to employ. Mechanically, there are no problems using fare classes. A passenger wanting to purchase a Q-class ticket from San Francisco through Chicago and on to Fargo can purchase the ticket if there's Q-class inventory on both of the flight legs. If not, the passenger is out of luck. As long as a method such as Littlewood's rule is generating nested fare-class levels for all of the flight legs in the network, actually performing the check on inventory is straightforward (see Figure 5.1).

The complications arise in the actual application of Littlewood's rule. When all passengers get on a flight at one city and get off at another, Littlewood's rule works without problem. All it needs is a list of fare classes, fares, and forecasts of demand for each class. With people starting their flights in many different locations and changing planes along the way, however, the fare associated with each class is no longer clear. An airline offering service to Fargo through a hub in Chicago might find passengers originating in Chicago paying $200 for a Q-class ticket while passengers originating in New York are paying $400 and passengers from San Francisco are paying $600. All three Q-class tickets draw upon the

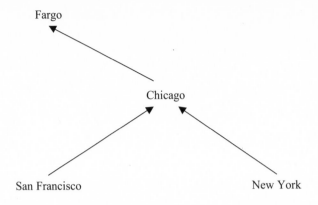

Flight Leg	Q Class Inventory
Chicago – Fargo	6
San Francisco – Chicago	8
New York – Chicago	0

Figure 5.1 Each flight leg maintains its own bucket of Q-class inventory. To book a flight in Q class, there must be Q-class inventory on all fight legs in an itinerary. With the inventory levels shown, a passenger can book a Q-class ticket on the Chicago–Fargo flight leg, the San Francisco–Chicago flight leg, or the San Francisco–Chicago–Fargo itinerary. The New York–Chicago flight leg and the New York–Chicago–Fargo itinerary are both closed to the sale of Q-class tickets since there are no Q-class tickets in inventory on the New York–Chicago flight leg.

same bucket of Q-class inventory. So what Q-class fare should be used in Littlewood's rule when calculating nested inventory levels on the Chicago–Fargo flight leg?

Difficult as the problem of choosing a fare is, there's a related and even more challenging problem. If someone comes along who wants to pay $600 for a ticket from San Francisco to Fargo, it seems like a good choice for the airline compared to someone paying $400 for traveling from New York or $200 from Chicago. However, the passengers flying from New York and San Francisco use *two* flight legs, not just one. These passengers are using more seats than the Chicago–Fargo passenger, and this needs to be accounted for. If the San Francisco–Chicago flight feeds a large number of passengers into Chicago that pay $1,000 to fly to another location, say, London, the *opportunity cost* of using the San Francisco–Chicago flight leg to fly someone to Fargo may make it an unprofitable decision (see Figure 5.2). Somehow, the opportunity costs on other flight legs need to be accounted for in the decision-making process.

Pseudofares were developed to address the twin problems of accounting for the opportunity costs of other flight legs and assigning a good fare to each fare class (details about how pseudofares are calculated and used are contained in the appendix). For each fare class and each itinerary, the fare is apportioned to the different flight legs in the itinerary using opportunity costs calculated from

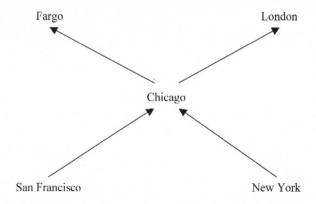

Itinerary	Passenger Demand	Q Fare
Chicago – Fargo	1	$200
New York – Chicago – Fargo	1	$400
San Francisco – Chicago – Fargo	1	$600
San Francisco – Chicago – London	1	$1000

Figure 5.2 One seat is available on each of the four flight legs. Ignoring the network and looking just at the demand that uses the Chicago–Fargo flight leg, the $600 San Francisco–Chicago–Fargo passenger appears to generate the most revenue, suggesting that this is a good passenger to accept. However, the passenger displaces the high-paying San Francisco–Chicago–London passenger. The best solution for the entire flight network is to choose the New York–Chicago–Fargo passenger and the San Francisco–Chicago–London passenger.

solving a mathematical model similar to that of the CFO problem. The resultant pseudofares on each flight leg are then averaged within each fare class to arrive at a single fare. This average fare is used by Littlewood's rule as the fare-class fare.

Pseudofares work surprisingly well within the restrictions imposed by using buckets of fare-class inventory on each flight leg. The problem is that these buckets are simply a poor way to manage inventory for a flight network in which people fly on multiple flight legs. No matter what is done to improve the underlying mathematical models that determine nested fare-class levels, fare-class buckets simply aren't capable of selecting the most profitable passengers (see Figure 5.3). Pseudofares are a good band-aid, but there are better ways of managing inventory.

One ingeniously simple idea is to work directly with the opportunity costs rather than use them to create pseudofares. If an opportunity cost represents how much each seat on a flight leg is expected to make, why not part with a ticket only when it generates at least as much revenue as the opportunity cost? For example, in Figure 5.4, the opportunity cost of the New York–Chicago–Fargo itinerary is $0 + $400 = $400, while the price of a Q-class ticket is $400. Since the revenue is at least as large as the opportunity cost, the airline should sell the ticket. On the

Los Angeles ⟶ Dallas ⟶ Miami

Itinerary	Passenger Demand	Q Fare
Los Angeles – Dallas	1	$300
Dallas – Miami	1	$400
Los Angeles – Dallas – Miami	1	$600

Figure 5.3 Assuming there is one seat remaining on both the Los Angeles–Dallas flight leg and the Dallas–Miami flight leg, the carrier would like to accept the two passengers wanting to fly on each individual flight leg for a revenue of $300 + $400. However, opening Q class on each of these flight legs also leaves Q class open for the $600 passenger wanting to fly on the Los Angeles–Dallas–Miami itinerary. It's impossible to simultaneously block the Los Angeles–Dallas–Miami passenger while making Q class available to the other two passengers using fare class inventory buckets.

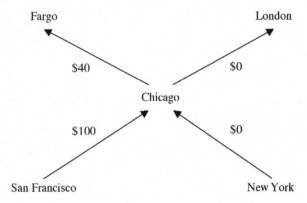

Itinerary	Passenger Demand	Q Fare	Opportunity Cost	
Chicago – Fargo	1	$200	$400	
New York – Chicago – Fargo	1	$400	$400	($0 + $400)
San Francisco – Chicago – Fargo	1	$600	$1400	($1000 + $400)
San Francisco – Chicago – London	1	$1000	$1000	($1000 + $0)

Figure 5.4 The opportunity cost method allows only the most profitable set of passengers to buy tickets, since only the New York–Chicago–Fargo passenger and the San Francisco–Chicago–London passenger have fares at least as large as the opportunity cost of the flight legs used. Observe that the method does not do away with fare classes, but does eliminate fare class buckets at each leg for deciding when a fare class is or isn't available for sale.

other hand, the opportunity cost of the San Francisco–Chicago–Fargo flight is $1000 + $400 = $1400, which is far in excess of the Q-class price of only $600. In this case, the ticket should not be sold. The opportunity cost method overcomes the problem described in Figure 5.3 (see Figure 5.5).

Los Angeles ⟶ Dallas ⟶ Miami
$300 $400

Itinerary	Passenger Demand	Q Fare	Opportunity Cost
Los Angeles – Dallas	1	$300	$300
Dallas – Miami	1	$400	$400
Los Angeles – Dallas – Miami	1	$600	$700 ($300 + $400)

Figure 5.5 Using opportunity costs, the Los Angeles–Dallas passenger and the Dallas–Miami passenger are able to buy tickets since the Q-class fare is at least as large as the opportunity cost. The Los Angeles–Dallas–Miami passenger is not able to buy a ticket since the Q-class fare is less than the opportunity cost. The opportunity costs were calculated using a mathematical model similar to that of the CFO problem.

Opportunity costs are an excellent way of controlling what's for sale, and they are capable of generating revenues that fare class bucket methods simply can't produce. This alone is enough to warrant their use, but opportunity costs offer other advantages as well. By providing a value for each unit of inventory, opportunity costs can be used elsewhere in airline operations. For example, a flight with a high opportunity cost indicates that inventory on the flight leg is extremely valuable, so that an airline might want to introduce a larger capacity plane. Further, the method is easy to use and intuitive to understand, although there always remain questions about the mathematics used to calculate the opportunity costs. Still, while the use of opportunity costs to manage the sale of inventory is gaining popularity in the travel industry and appears likely to become dominant, acceptance has been slow.

Airlines have been setting fare class inventory levels for such a long time that it's difficult for people who work with them to let go. A revenue management analyst's life revolves around seats, not opportunity costs. Organizational issues also pose a challenge as revenue management departments change from a focus on flight legs to a focus on network traffic.

However, the biggest reason for slow acceptance has to do with reservations systems and the distribution network they support. Reservations systems were designed around the concept of fare classes—how many seats are available on any flight leg in any fare class. Fare classes are now part of the infrastructure supporting the way airlines operate. Changing from fare classes to another means of managing what's available for sale and at what price is a very expensive proposition.

Even if an airline changes its own reservations system to work with opportunity costs, it still must deal with the rest of the travel industry. The primary means by which a carrier communicates what it has available for sale to third parties—alliance partners and travel agents, for example—is still through fare class availability on each flight leg. Again, all these limitations can and are changing, but the process will take years. Change will occur, however. The limitations imposed by fare class buckets aren't just a nuisance, they're costing the airlines money.

* * *

Whether an airline uses nested fare class buckets or the opportunity cost method to determine what fare classes are for sale, the net result for passengers is the price we find when trying to buy a ticket. How does an understanding of the science of airline ticket pricing explain some of the price changes we see?

Orbitz is one of the more interesting travel Web sites in that it promotes comparison shopping. Provided with the basic information requested by all travel sites—origin, destination, dates of travel, and approximate times of departure—Orbitz returns a matrix of possibilities. The columns in the matrix are labeled with the names of different airlines that service the origin and destination cities, while the rows list the number of intervening stops—zero, one, and two or more. Within each cell of the matrix is the lowest price for the given carrier and the given number of stops, within the specified time range on the desired dates of travel. Airlines are listed from left to right in the order of least expensive to most expensive.

Looking at the different prices presented by Orbitz, a variety of things stands out. More often than not, the most expensive carrier offers a price that is two or three times as much as the least expensive carrier. Is this because of poor, uncompetitive pricing? In fact, it may be because the carrier with the expensive flight has determined that the inventory is too valuable from the perspective of the overall flight network to part with it for any less. While we only see a price from point A to point B, the carrier sees an impact on its network of seats. To us as consumers, the large discrepancy simply seems out of touch with the market.

Network inventory valuation is only one of the factors impacting the price we see, and it is actually one of the more subtle elements. More straightforward factors involve a plane filling up faster or slower than expected. The easiest way to observe price changes due to this phenomenon is to visit the Web site of a carrier with many flights serving a particular city pair. The Friday night flight at five o'clock may be considerably more expensive than the flight at eleven o'clock in the morning. With high likelihood, both flights offer the same fare classes at the same prices. But since the five o'clock flight is more popular with business travelers, it fills up more quickly. With fewer seats available, the remaining seats become more valuable and the lower-priced fare classes are closed. A good revenue management system, knowing what to expect in the future, will actually shut the lower-priced fare classes even before the plane starts to fill up.

Group bookings also impact seat availability and therefore price. If a large party calls an airline months in advance to book 50 seats on a flight, the remaining seats become that much more valuable and lower-value inventory classes are shut down. The price offered to the group is itself established by taking into account the regular passengers the party is expected to displace in the future.

Of course, all of these different effects can't be easily separated. When a group makes a purchase, the impact is measured by the effects on the network. Sometimes the many interacting effects occurring within the network make it difficult to track down why fare classes are being shut down and prices are increasing on a particular itinerary, but often it's not difficult to track down the root cause. A big jump in the observed price on a flight itinerary is usually due to one of its flight legs receiving a large number of bookings.

The price we see is also impacted by how often an airline relies on short-term changes and how often it makes long-term changes by running the algorithms in its revenue management system. For example, when an airline uses nested fare-class control, short-term changes consist of simply updating how many seats are available for each fare class. If a carrier has 100 seats available for Y class, 70 seats available for Q class, and it sells a Q-class ticket, then after the transaction, it has 99 and 69 nested seats available for the two fare classes, respectively. If enough Q-class tickets are sold in this way, Q class can be closed for sale and the observed price of a ticket goes up. The reservations system handles this counting without any intervention on the part of a revenue management system apart from setting the initial inventory levels.

However, the levels of 100 and 70 may have been established when the airline first made the flight available for sale, typically several months to a year before its departure. As time moves forward, the airline gets to observe how many passengers are actually buying tickets. As a result, it may want to revise its demand estimates. Thus, airlines periodically reforecast, rerun an optimization algorithm to calculate new opportunity costs, and apply Littlewood's rule to establish new nested fare class limits. This periodic, long-term process may bring about abrupt changes to nested fare class limits and, as a result, lead to large observed increases or decreases in price. The opportunity cost method works similarly, but doesn't convert the opportunity costs to nested fare class limits.

At most airlines, reforecasting and reoptimization occur up to a few dozen times in the life of a flight network, with most of the activity occurring closer to the date of departure. A typical schedule is shown in Table 5.1. The calculations may take many hours and typically occur overnight, with availability limits changing in the morning. As computers have grown more powerful and airlines more sophisticated, the frequency has increased. Actually measuring the frequency becomes difficult since carriers may solve one part of the problem more often than others. Also, some carriers use "on demand" forecasting and optimization, letting booking activity trigger what gets done rather than scheduling tasks in an orderly, periodic way. Nonetheless, the numbers are staggering. An international carrier with a fleet of 100 aircraft using a state-of-the-art system

Table 5.1 Typical reforecasting and reoptimization schedule for a flight network, measured in days prior to the departure date.

360	180	90	45	30	25	21	17	14	11	7	5	4	3	2	1

might keep over a quarter of a billion forecasts on hand at any given time, generating new ones at a rate of 5 million per hour—approximately 1400 per second. These forecasts in turn translate into some 25 million opportunity costs that are used to control the sale of inventory.

Behind the scenes, the mathematics continues to grow more sophisticated as well. While Littlewood's rule still makes its presence felt and is arguably one of the simplest mathematical methods to ever display such longevity, it is increasingly being supplanted by methods that don't assume that low-paying passengers arrive before their high-paying counterparts. Instead, passengers are modeled as arriving randomly, at different rates and at different times. The mathematical theory underlying these alternative methods has been well understood for over a half century, but the application of this theory to airline inventory control has primarily occurred in the last 15 years. Methods for calculating network opportunity costs have also gone through a series of advances during the same period. Again, the underlying mathematical theory is not what's new; rather, it's the application of this theory to airline inventory control problems. As these mathematical advances find their way into use, it affects the prices that we see.

From an abstract, mathematical perspective, whether looking at individual flight legs or an entire network, revenue management is a game of picking and choosing, and we are "only marbles in a bag." Airlines invest in planes, infrastructure, and people, then wait to see who's going to show up to buy tickets. Like turning valves, carriers open and close fare classes to control who does and who doesn't get on the plane. The goal of the game is clear: given who you think is going to show up, try to pick the passengers that will give the highest total revenue. Through analysis and experimentation, airlines have developed some extremely powerful tools to achieve this objective.

The big question is who's going to show up. Even for a complicated flight network, if we knew all the passengers who wanted to fly and what they were willing to pay, we could pick through the many combinations and find the collection of passengers that would generate the most revenue. In a formal mathematical sense, this problem of picking passengers is hard, but for all practical purposes, it's not nearly as difficult as dealing with the uncertainty of who we expect to show up. The fact that so much effort is placed on determining nested fare class levels or opportunity costs derives from uncertainty in who's coming and when they're expected to arrive. Without uncertainty we could dispense with most of the mathematical machinery and replace it with automated email informing potential passengers who had been selected and who, regrettably, there would be no space for.

As it is, uncertainty is pervasive. How do we deal with uncertainty, and exactly how does it impact revenue? If we don't know with a high degree of certainty who's coming, doesn't this negate the value of scientific pricing? Like so many questions, the answer is surprising.

Note

- Readers interested in more details about the methods used by network airlines can find good summaries in Boyd and Bilegan (2003), McGill and van Ryzin (1999), and Talluri and van Ryzin (2004). These works also address the history of the development of airline revenue management, though, since many of the ideas originated in industry, it is often difficult to pinpoint exactly when and where they began.

CHAPTER 6

Hold Me, Darlin'

The 2005 World Poker Tour Championship took place over a stretch of seven days in late April at the Bellagio casino in Las Vegas. With 452 entrants putting up $25,000 a piece for the right to play, the total purse for the competition came to over $11 million. The winner would take home almost $3 million, one of the largest prizes in the history of tournament poker.

The game was No Limit Texas Hold 'em, the most popular poker game played today. Folklore has it that Hold 'em began in the back rooms of Texas in the early twentieth century, and it generated such heart stopping play that the original name was "Hold me, Darlin'," evoking images of a hardened, overweight man grasping at his chest as he comes to terms with the will of fate.

Seated at the final table were a 26 year old named Tuan Le and a retired Englishman named Paul Maxfield. In keeping with World Poker Tour tradition, with only two players remaining, the first-place prize money was brought into the room with great ceremony and dumped in a pile of small bills on the table. The trophy was there as well, of course, but poker players hold no illusions about why they play. Football, basketball, baseball; they all involve vast sums of money. But who other than poker players so brazenly celebrate the money as much as the victory?

After an exhausting battle, Le held almost three-quarters of the 22 million in chips on the table—a commanding lead, but far from making him a guaranteed winner. With 750 thousand chips in the pot from antes and blinds, it was Maxfield's turn to bet first. Glancing at his cards, he reached into his stack and pushed forward another two million. Le looked at his own cards, the ace and four of diamonds. Were his cards good enough to call, or should he fold? Perhaps he should raise. If so, by how much?

* * *

There are many times when we'd like to know what will happen in the future, but, the world being what it is, we need to settle for our best guess. The important

Sidebar 6.1: No Limit Hold 'em

Hold 'em proceeds with each player at the table receiving two hidden cards followed by five community cards—shared cards that are placed face-up in the middle of the table. Players bet after they receive their first two cards, after which three community cards known as the *flop* are turned over simultaneously and another bet is made. The fourth community card, known as the *turn*, is dealt, followed by another bet. The final bet comes after the fifth and final community card, known as the *river*, is dealt. Players then reveal their best five-card hand from the five on the board and their two hidden cards.

In no-limit poker, there are no restrictions on the amount a player can bet. Should an opponent put more chips into the pot than a player has in front of him, the player may call by pushing in what chips he has and declaring "all-in." Should the player with fewer chips lose, his chips are gone and he is out of the game. If he should win, he only wins an amount equal to what he put in the pot. In tournament poker, play continues until one person holds all of the chips. A player's finishing place is determined by when his chips are lost relative to the other players, and prize payments are made according to a predetermined payout schedule. Thus, a player holding "10 million in chips" is actually holding 10 million *units* in chips, not dollars.

question is how we go about making such a guess. Often we rely on our common sense, and our common sense serves us well. We aren't afraid of car travel because experience has taught us that the likelihood of being seriously injured is small enough to offset the convenience of driving (although in reality, driving a car is more dangerous than most of us admit). Common sense tells us that buying a diversified portfolio of stocks and holding them for an extended period is a good investment, and, more often than not, it is.

Yet as much as we rely on our common sense, it can also fool us. The Secretary Problem described in Chapter 4 is a good example of how our common sense can fail. It seems impossible that in a trillion numbers we know nothing about we can find the largest more than a third of the time. Yet this is exactly what a little mathematical analysis tells us.

Another example in which our common sense can fail is captured in the forecasts depicted in Figure 6.1. The diamond line shows the actual number of packages arriving at a small shipping facility on Fridays over an 18 week period. No forecasts are shown for the first six weeks, but thereafter, a forecast is made one week in advance using two different forecasting methods. For example, the forecast of 60 by forecast method 2 in week 16 was made during week 15 using all the observations of package arrivals on Fridays in the first 15 weeks.

Forecast method 1 appears to be performing reasonably well. It is tracking the general pattern of demand, moving up when demand moves up, and moving down when demand moves down. On the other hand, forecast method 2 appears

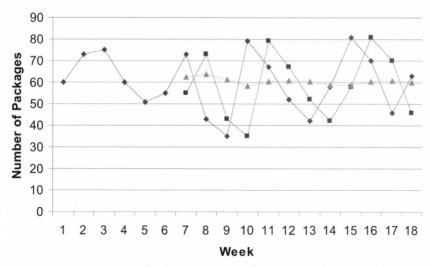

Figure 6.1 The actual number of packages arriving at a warehouse versus two different forecasts made one week in advance. Forecast method 1 uses the actual number of packages observed this week as next week's forecast. Forecast method 2 uses the average number of packages observed this week and all preceding weeks as next week's forecast. The average error made by forecast method 1 is 40 percent higher than that of forecast method 2.

to be sluggish, reacting to history, but not picking up changes as needed. In actuality, if we take a look at the total error made by forecast method 1, it's 40 percent larger than the error made by forecast method 2.

This simple example carries a multitude of important lessons. First, it's necessary to be very careful when evaluating forecasts. Not only can we deceive ourselves, but beauty is in the eye of the beholder. Many times I've been faced with two people looking at identical figures comparing forecasts to actual observations, one thinking the forecasts are outstanding and the other thinking they're miserable. In the example of Figure 6.1, we have a clear measure for determining that forecast method 2 is better than forecast method 1. But if we look just at the forecasts generated by forecast method 2 compared to the actual observations, without the benefit of the reference frame provided by forecast method 1, are the forecasts good or bad? This is a very difficult question, because without an agreed upon frame of reference, each person chooses his own. It's not that two people are looking at a single glass, one seeing it as half full and the other seeing it as half empty. It's more that one person is looking at a glass that's actually full and the other is looking at a glass that's actually empty. They've just chosen to look at different glasses.

Not uncommonly, people have expectations about how good forecasts should be without any foundation for their reasoning. Suppose, for example, someone is looking for a computer system that can forecast the sum of the roll of two fair, six-sided dice, with the requirement that the error of the forecasting system be within two at least 75 percent of the time. It's not difficult to see that this is simply impossible. The best solution is to forecast seven on every roll, in which case the roll will be within two roughly 66.7 percent of the time. Not only is forecasting incapable achieving the desired goal, but we can also go further and quantify how far off we expect to be.

Forecasting events such as how many packages will show up on a given day is in some ways similar to forecasting the sum of two dice, but there are differences. When dice are rolled, the outcome is random in a way that we completely understand. We accept and quantify this randomness. When the number of packages that shows up isn't what we expect, we begin to question ourselves. Have we really captured all the various factors that can influence demand? The overall decline in local wages may have people watching their expenditures more carefully than in the past. Perhaps the end of the recent street construction is making it easier for people to get to the store. Maybe people are scared to venture out because the moon is full.

A litany of factors can impact forecasts. Finding the relevant factors and incorporating them in forecasting methods can improve forecasts and is an important part of forecasting. Seasonality is a commonly used factor. The number of packages shipped before Christmas is understandably higher than that at other times of the year. In the back of our minds, we'd like to find all the contributing factors and include them when making our forecasts. If we keep finding more and more of the important factors, we can make our forecasts better and better. In part, it is this belief that helps fuel the notion that we can pick a forecasting criterion—"90 percent of all forecasts must be within 10 percent of actual observations"—and drive toward that goal until we achieve it.

A little thought, however, makes it clear that we are only deceiving ourselves with such expectations. Even if we knew all the factors that influence demand, how do we put them all together? If the temperature, the day of the week, the price of the competitor across the street, and the phase of the moon all impact demand, and if it's 48 degrees on a Thursday, my competitor is willing to ship a small package to Madison, Wisconsin, for $3.80, and the moon is a waxing crescent, how many people do I expect to show up today? Complicating the issue is that forecasting models using this information—and good models of this nature do exist—find it more and more difficult to estimate the impact of the factors with each additional factor that's incorporated. The fact is that no matter what we include, there will always be some degree of randomness that we can't account for. Uncle Jim may decide to ship all those old boxes of pictures to Aunt Sally simply because he woke up in the morning with a bit more energy than usual.

Accepting this inherent randomness is a liberating experience. Rather than engaging in a never-ending struggle to grab the brass ring that always seems just out of reach, we can begin to deal with randomness for what it is—a natural part of the world we live in. Those who embrace randomness understand that it's not necessary to know exactly what's going to happen in order to take advantage of the future. But dealing with randomness in a rational fashion is a prerequisite.

When Edward Thorp published his classic *Beat the Dealer: A Winning Strategy for the Game of Twenty-One* in 1962, the world of casino gambling changed forever. Based in part on an article published in the *Proceedings of the National Academy of Sciences* two years earlier entitled "*A Favorable Strategy for Twenty-One*," the book developed an approach to playing the card game more commonly known as blackjack.

The important word in both of Thorp's titles is *strategy*. In football, our strategy might be built around a running game or a passing game. In basketball, our strategy might be getting the ball to our 7'8" center and letting him work inside to the basket, or it might be moving the ball around for an open shot. Whatever we do, we need a strategy that underlies how we make our decisions. Any sports coach will tell us that without a strategy, without a sense of what our strengths are and how we will build on those strengths, we will fail. We need to think carefully about what strategy we choose and then execute that strategy, not being sidetracked by yesterday's random events. A bad game is a bad game. Certainly, if our strategy isn't working, we should re-evaluate it and change it if necessary. But the process needs to be thoughtful, logical, and consistent.

In blackjack, a static strategy consists of telling us what to do if the dealer shows card X and the player holds cards Y and Z. Do we take another card, stand, or, if the option is available, split the pair or double-down? Strategies of this type can be found in tabular form on the Internet. They're easy to memorize, and many casinos will let you bring printed copies to the table since there are no

---

Sidebar 6.2: Blackjack

In blackjack, the player's objective is to have his cards total closer to 21 than the dealer without going over 21. Face cards are worth 10 and aces are worth one or 11 at the player's discretion. Both the player and the dealer start with two cards. The dealer shows one card, and the player must complete his turn before the dealer completes hers. In this way, if the player makes a total greater than 21, he loses no matter what eventually happens to the dealer. This is the source of the house advantage.

To partially offset this advantage, players receive different treatment than the dealer. For example, when a player reaches a total of 21 with exactly two cards, the house normally pays one and a half times the player's wager rather than the usual one-for-one. Players also have the option of splitting pairs. When dealt a pair, the player may form two new hands by separating the pair and receiving a new card to go with each. In doing so, the player must place a wager equal to the original wager on each hand. Another option is that of doubling-down, in which a player may choose to double the original wager while limiting himself to taking one and only one more card. This is often advantageous when the player is dealt two cards totaling 10 or 11. Many small variations in the game exist depending upon the casino.

known static strategies that provide the player an advantage over the house. Such strategies can reduce the house advantage to a very small percentage, but the house still retains an advantage.

Thorp's insight was that the player's probability of winning, and his strategy, change as certain cards are revealed in the course of play. Recognizing this, Thorp was able to show that by "counting" cards—keeping an estimate of the deck's favorability—a player could actually expect to win money from the casino.

Randomness is part of the game of blackjack. Sitting at the table with a total of 14 while the dealer shows an eight doesn't leave us feeling particularly confident that we're going to win. We gather our courage, knowing that the right thing to do is to take another card, and ask the dealer for a "hit." There's nothing we could have done to change the cards we were dealt, nor is there anything we can do to affect the card that's coming.

Nor do good blackjack players who count cards really care. They understand that if they make the right move in each situation, they're going to make money. Maybe not on this hand, but as they play more and more hands, their winning is a virtual certainty—just as certain as the fact that the novice sitting to their left is going to lose. To the good player, blackjack isn't exciting, but it won't lead to sleepless nights at home wondering what went wrong with the cards.

The key to winning is to play a good strategy day in and day out. Figure 6.2 shows the winnings for a player using a strategy that gives the house a 6 percent advantage, typical for a player who knows the game but has no real sense of the probabilities involved. Due to the randomness inherent in the game, the player experiences periods of winning and losing, but the net trend is downward.

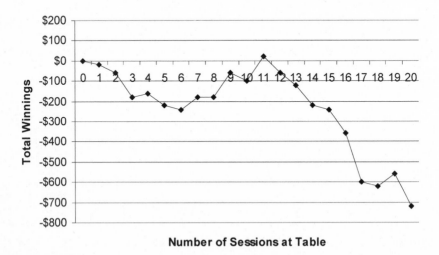

Figure 6.2 Example of the total winnings of a player after each of 20 sessions at the blackjack table. Each session consists of 60 hands played for $10. The player is using a strategy in which the house has a 6 percent advantage. The expected loss is $720.

Figure 6.3 shows similar results, but for a player using a strategy that yields a 0.2 percent edge over the house. This probability is in line with what a good card counter can hope to achieve. With such a tiny advantage, the winnings tend to move up and down with more frequency, and they aren't as pronounced as the losses associated with a 6 percent house advantage. Still, the trend is upward. Card counters can increase their expected winnings by betting $100 or $500 on each hand instead of only $10, but however you look at it, counting cards is a

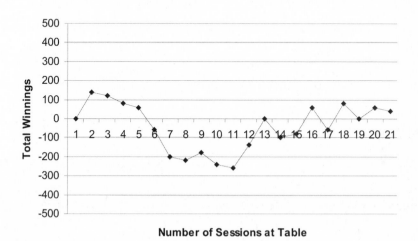

Figure 6.3 Example of the total winnings of a player after each of 20 sessions at the blackjack table. Each session consists of 60 hands played for $10. The player is using a strategy with a 0.2 percent advantage over the house. The expected win is $24.

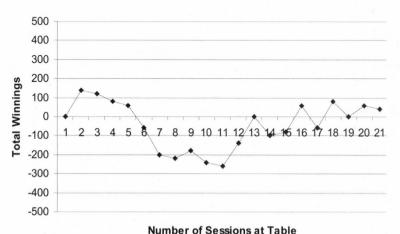

Figure 6.4 Example of the total winnings of a player after each of 20 sessions at the blackjack table. Each session consists of 60 hands played for $10. The player is using a strategy with a 6 percent advantage over the house. The expected win is $720.

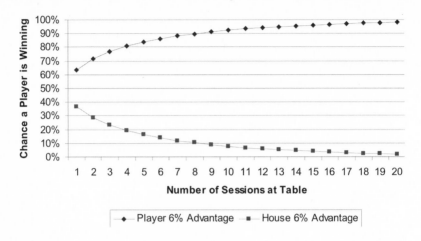

Figure 6.5 The chance that a player is winning after each of 20 sessions at the blackjack table. Each session consists of 60 hands played for $10.

tough life. For reference, although this advantage could never be achieved by any legal means, Figure 6.4 depicts the winnings of a player using a strategy with a hypothetical 6 percent advantage over the house.

To get some sense of the risk involved, we might want to determine the chance that we're ahead after a combined number of sessions. Figure 6.5 shows the striking difference between a 6 percent player advantage and a 6 percent house advantage. After only eight sessions at the table, a 6 percent player advantage translates into a 90 percent chance of walking away a winner. By the end of all 20 sessions, the chance of winning is all but certain. The fact that our chances of winning get better the more we play should come as no surprise. What's surprising is how even small differences add up so quickly.

I don't find myself in Las Vegas too often, but when there, I enjoy a few hands of blackjack. I don't play for the money since it's too much work for an evening of entertainment, and because even a casual effort at counting cards is likely to get noticed by the pit bosses. Using a good basic strategy allows me to enjoy a free drink and kibitz with those at the table without fear of losing too much money.

It's easy to pick out the good and the poor players based on how they play certain hands. Splitting tens is never a good idea. Splitting eights is always the proper play. Yet most inexperienced players will do just the opposite until coached otherwise. Helping someone learn to play better is always fun, especially when you've spent the last 15 minutes learning that she's from Tulsa, she really enjoyed the show at the Mandalay Bay last night, and her youngest son didn't finish college but is a good boy and will make something of himself someday. Everyone at the table is in it together, with the common purpose of beating the dealer. With a little luck, the dealer will turn out to be friendly and join the players, offering encouraging banter and reminding us that we're really playing against the house. Some dealers, when faced with an extremely poor player, will even offer help. Whenever I've observed this, it's always been good advice.

The biggest lesson I've learned from my time at the blackjack table is just how much people believe in luck—luck to the exclusion of rational thought. Some people have no sense of what probability is, while others grasp the basic concept but have never managed to internalize what it really means. They remember how to calculate simple probabilities, but if they choose to stand on 12 when the dealer is showing a nine (a horrible decision), what difference does it make? Winning isn't about strategy, it's about finding the right table at the right time. It's about going with your gut instinct.

On one occasion, I was sitting next to a woman who was playing very conservatively. If she held a hand totaling 12 or more, meaning that the next card could put her over 21 and out of the game, she would stand. Her conservative play, however, didn't mean she displayed a conservative demeanor. The life of the table, she would cheer loudly when she won a hand, and even more loudly when the dealer went bust. It was all in good spirit, and even the dealer was having a good time.

At one point she held a total of ten with the dealer showing a six, a very favorable position for the player with about a 65 percent chance of winning using the optimal strategy of taking one card and standing. She knew she could double-down (see Sidebar 6.2), but was having trouble making a decision. The right play was absolutely clear, and I couldn't understand why she was having trouble doubling her bet when the odds were so strongly in her favor. The only conclusion I could reach was that she didn't have a sense of what an advantageous position she was in. I explained her good fortune to her, and in the spirit of the moment, vigorously encouraged her to throw in the extra chips. Still, she hesitated, and I couldn't understand if it was because she didn't believe me, if she was reasoning the problem on her own, or if she was drawing upon a sixth sense in an effort to predict what was going to happen. In the end, she took my advice, but the cards went the wrong way. She drew a four to make 14, and lost when the dealer made 18. With a look I will never forget, she returned to playing. But it was clear she wasn't going to take my advice anymore that evening, nor, I suspect, in the future if our paths ever crossed again.

Without an appreciation of how to deal with randomness, my table companion was destined to lose. Maybe not that evening, but as certainly as dropping a hammer means it will fall to the ground, I can guarantee that if she's playing regularly the way she did that night, she's lost money at blackjack, and a lot of it.

My table companion can change, as can anyone who deals with randomness. Whether predicting how many packages will arrive at a shipping facility, ascertaining if it's safe to ride a motorcycle, or trying to close a sale, randomness lurks. Accepting randomness and dealing with it intelligently, we develop an advantage most people in the world will never discover, and as a side benefit, we make those day-to-day fluctuations in our lives a lot less painful to deal with.

* * *

With only two players, Tuan Le's ace and four of diamonds represented a good starting hand in Hold 'em. Knowing nothing about his opponent's hand, Le had a 57 percent chance of winning. Recognizing that he held a good hand in turn allowed Le to intelligently evaluate the options in front of him. With Maxfield raising by two million in chips, Le had to suspect Maxfield held something worth betting on. Yet late in the game when the antes and blinds are high and a player is running low on chips, he is often forced to play weaker hands than he would in other situations. In order to improve his chances of winning by giving Maxfield the option of folding, Le chose to raise with all his chips. Maxfield responded with his own all-in, and the race was on. Maxfield would either take the chip lead, or be out in second place.

With no decisions left to be made, the players stood and flipped their cards over, showing them to all present. Maxfield held the king of spades and the eight of diamonds. With the high card and two cards of the same suit, Le was in good position. Maxfield's only real chance to win was to pair either his king or his eight. Even if he did, he would have to hope that Le didn't make a pair of aces or catch a flush. There were other possibilities, but Le's chance of winning, which had risen to 62 percent once Maxfield's hand was revealed, largely depended on the likelihood that Maxfield wouldn't make a pair. Unable to look, Maxfield turned and walked away from the table. Le could only watch and wait to see what the next five cards would bring.

The flop brought the nine of hearts, ten of clubs, and five of spades, an excellent result for Le, lifting his chance of winning to 72 percent. Maxfield now had only two cards coming to draw a king or an eight.

The next card brought a minor surprise. When the seven of spades hit the board, Maxfield found himself looking at the possibility of a straight. With the eight in his hand and a seven, nine, and ten on the board, Maxfield could now win if the next card was a six or a jack as well as a king or an eight. What had been a remote possibility only one card earlier now gave Maxfield hope. Had the turn card been something innocuous, say, the four of spades instead of the seven, Le would have been looking at an 86 percent chance of winning while waiting for the last card to be revealed. As it was, Le's chances dipped to 68 percent—still good odds for winning the tournament.

The dealer tapped her hand on the table, discarded the top card from the deck, and turned over . . . a king. Maxfield threw up his hands in an expression of joy and relief. He was now the new chip leader. Le, having been the favorite to win the hand until the very last card, gathered himself and sat back down.

* * *

Poker players have many names for an especially unlucky turn of events—bad beat, suck out, and other more expressive terms that arise in the heat of the moment. There isn't a poker player on earth who doesn't have a story about how the gods of chance frowned upon him. Good poker players know, however, that while they can't control the outcome of the cards, they can use science to help them make the right play at the right time. Tuan Le made such a play. Knowing the probabilities and taking into account all information at his disposal, Le found himself in a good position to win the tournament until he met with an unfortunate twist of fate. Even so, he hadn't lost the tournament, he'd only lost one hand. There were many more to be played.

How do we know when we're playing a good strategy? In some instances, when the problem is simple enough, we can mathematically prove that a particular strategy is not only good but optimal. As we saw in Chapter 4, Littlewood's rule is optimal for choosing nested fare class levels for two fare classes when all the low-paying passengers arrive before the high-paying passengers. No other strategy provides higher revenues on average. However, when problems grow more complicated, we need to rely on simulations.

The quality of a particular blackjack strategy is determined by letting a computer play the strategy tens or hundreds of thousands of times, keeping a record of how much money a player wins or loses. Thanks to the properties of random events, properties that we understand from a lengthy history of theoretical work, we can develop very accurate statistics about any given strategy.

A similar process is used for evaluating pricing strategies. Figure 6.6 shows how much money an airline makes over a 30 week period on a particular flight using three different strategies. The simulation operates by generating a random number of passengers for each of two different fare classes. These passengers are then lined up in some order and "arrive" to make a purchase. If no ticket in their desired fare class is available, they walk away empty handed. Otherwise, the transaction is recorded, the airline receives the price of the ticket, and a seat is taken out of the inventory.

As can be seen in Figure 6.6, no single strategy generates the most revenue in every single week. A casual investigation to determine which strategy is best could easily lead to an incorrect conclusion. Strategy 1 is clearly inferior, but at times it is the best performer. Looking at the entire 30 week period, we see that Strategy 2 shows revenues that are 21.4 percent better than Strategy 1, with Strategy 3 yielding 24.6 percent greater revenues than Strategy 1. The additional 3.2 percent revenue comes from nothing more than the simple calculations associated with Littlewood's rule, and staying true to this strategy.

The actual simulations used to evaluate airline pricing strategies are significantly more complex than those described here, and may include passengers that cancel, more realistic arrival orders than low-paying before high-paying passengers, and many other eventualities. However, the basic approach of letting the computer generate streams of arriving passengers, observing how they book, tallying the revenues, and repeating the process over and over remains unchanged.

Researchers have studied, modeled, analyzed, and simulated many different scenarios of how passengers arrive and mathematical strategies for deciding who

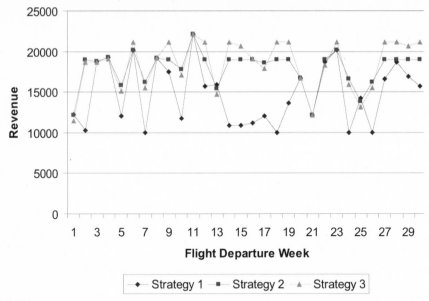

Figure 6.6 Example of the total revenue generated by an airline on a Wednesday morning flight from Philadelphia to Albany for 30 successive Wednesdays. The aircraft has 100 seats and two fare classes. Y class sells for $300 and Q class sells for $100. Y-class demand comes from a bell curve with a mean of 30 and a standard deviation of 10, while Q-class demand comes from a bell curve with a mean of 90 and a standard deviation of 40. Low-paying Q passengers arrive before high-paying Y passengers. Three different strategies are used:

> **Strategy 1:** No control. Let passengers purchase whatever ticket they desire as long as there is a seat available.
>
> **Strategy 2:** Mean control. With an average of 30 Y-class passengers expected to arrive, use nested fare class limits that restrict Q-class purchases to no more than 70.
>
> **Strategy 3:** Littlewood's rule. Use Littlewood's rule to calculate nested fare class limits that restrict Q-class purchases to no more than 63.

Over the 30 week period, Strategy 2 generates 21.4 percent more revenue than Strategy 1, and Strategy 3 generates 24.6 percent more revenue than Strategy 1.

to accept and who to turn away. While they have given rise to a wealth of research literature, far more than could be covered in a few brief pages, several important ideas stand out as central in driving the evolution of airline pricing.

Two ideas of special significance have to do with the interpretation of observed ticket-sales data. At first glance, it's hard to imagine there's anything worth considering. If the number of Q-class tickets sold over the last five weeks is 70, 70, 60, 70, and 50, then it seems reasonable to take their average of 64 as an estimate of the number of passengers we expect will want to buy Q-class tickets next week. But what if during the previous five weeks we never allowed ourselves to sell more than 70 seats in Q class, an event that is quite likely as we close fare classes? We may have gone many days or weeks during which passengers showed up wanting to buy a Q-class ticket but couldn't. For the weeks in which we observed the sale of 70 Q-class tickets, we know that there were at least that many passengers who wanted to buy a Q-class ticket, and probably more. The number 70 is an underestimate of Q-class demand for that week.

Accounting for historical underestimates of this type is known as *unconstraining*. Most industries, including the airlines, rely on mathematical techniques to reconstruct missing demand, while others try to keep record of people who attempt to purchase but don't. This latter technique is prone to error, especially with the ease of window shopping on the Internet. When a shopper shows up on a Web site and inquires about a hotel room only to find it isn't available, does that mean that if it had been available, the shopper would have reserved it? Or does it mean that the shopper was simply taking a look?

The second idea related to data interpretation is *buy-down*. Suppose our Wednesday morning flight never fills up and that we don't use any inventory controls, so that Y and Q class are always available for sale. In this case, there is nothing stopping us from observing all the Y- and Q-class demand and unconstraining isn't an issue. If people treat Y and Q class as truly differentiated products—such as dishwashers and microwave ovens—then looking at historical Y- and Q-class sales gives us reasonable estimates for the future. If, as we argued in Chapter 2, however, people frequently view Y and Q class as identical products differentiated only by price, then we're not observing the actual demand of customers by their willingness-to-pay. Many of the passengers we observe purchasing Q-class tickets might be willing to pay the Y-class price, but with Q class available, they are happy to buy down to the lower fare. We want to forecast what people are willing to pay, but what we actually observe is what they paid.

Mathematical methods exist to reconstruct passenger willingness-to-pay from the actual price paid. Buy-down is an especially vexing problem in that if it is not accounted for, it can severely reduce airline revenues. The problem is that buy-down is a one-way street. If airlines forecast fare classes as products, but even some of the people are purchasing based on the lowest available price, then the forecasting methods will always underestimate the number of people willing to pay the higher fare and overestimate the number of people willing to pay the lower fare. In turn, less seats will be protected for the higher-price class, making more available for the lower-price class, and the problem becomes self-perpetuating as new observations are used to forecast future demand.

Unconstraining and buy-down illustrate an important difference between forecasting for purposes of pricing and other types of forecasting. When forecasting the number of packages arriving at a shipping facility, what we see is what we can expect. Every week we observe the actual number of packages that arrive, and this is exactly what we are trying to forecast in the future. When forecasting for pricing purposes, however, what we see is *what we sold*, and what we want to know is *what we could have sold*—what the actual demand is at each price point.

The complications associated with unconstraining and buy-down are twofold. First, because historical sales observations aren't in line with what needs to be forecast, it is necessary to go through the process of correcting these observations in an effort to find out the actual demand at each price point. Second, once these corrected observations are used to forecast future demand at each price point, what we observe in the future is what's sold, not actual demand at each price point. The forecasts are estimating one thing, and the observations are of something else. Thus, a direct comparison of forecasts to observations is fundamentally wrong and leads to all sorts of problems when such comparisons are made, as they inevitably are. Good forecasts of demand at each price point can be generated, but evaluating their quality has an additional level of complexity not found in most forecasting applications.

A third idea of special significance is the effective use of competitor price information. Many people are surprised to learn that most airlines don't use competitor price information in their operational pricing activities. It just seems natural that with all the different Web sites displaying prices that this information must be used.

The problem is less one of concept than it is of practice. At any given instant, what is the price to fly from Cleveland to Las Vegas? The available price can vary depending on where you look, and it may change at any moment. This is less of a problem than actually obtaining available prices, since sites where this information is located are often difficult for competitors to access via electronic means. Airlines actively engage in regularly changing the details of their Web sites so that competitors' price-reading computer programs are confounded. Humans are sometimes employed to visit Web sites or make phone calls to obtain the best available price from a competitor, but this is expensive and impractical for most carriers in most markets.

The price of any fare class is available through industry organizations as discussed in Chapter 3, and airline pricing departments frequently use this information when setting their own prices. However, the prices don't reflect what fare classes are open or closed, and therefore what price is available in the market at any particular instant.

Operational pricing is successful without explicit use of competitor price information because this information is implicitly captured in observed sales. If a competitor closes fare classes and as a result doubles the available price, other carriers will see a rise in demand, close lower-price fare classes as they sell out, and see an overall rise in the available price they offer. If a competitor does this on a regular basis, demand forecasts will change and the lower-price fare classes will naturally be closed preemptively.

Changes in available price as a result of changes in observed sales are relatively slow compared to changes made in direct response to a competitor's price. If a competitor changes fare class availability and therefore price, a carrier might want to make an immediate adjustment to its own fare class availability. It's this perspective that keeps competitive price information one of the most hotly discussed issues in the revenue management community today. Still, even if the practical problem of gathering available price data is overcome, care must be exercised as this data is used. It's not difficult to imagine two carriers with pricing algorithms designed to undercut each other's price offerings by a small amount and, as a result, driving the available price as low as possible at the speed of light. Competitive price information is powerful, but it has the potential to lead to disaster.

A fourth idea of special significance is consumer choice. Consumer choice models take a slightly different approach to forecasting. Rather than asking, "How many passengers are willing to pay for a particular fare class on a particular morning flight from city A to city B next Tuesday," they ask, "If a person shows up looking to fly from city A to city B next Tuesday morning, and there are three different flight itineraries, and fare class Y is open on the first, M on the second, and Q on the third, how likely is it that the person will choose Y class on the first flight?"

The two questions are very different. In the first case, the goal is to come up with a number. Like forecasting how many packages we expect to arrive at a shipping facility, we can use what we've seen in the past or use other factors to make our prediction, but we're after a particular number. In the second case, we're seeking to understand how consumers react when faced with a particular set of options. Understanding how consumers make choices is an alluring proposition. After all, we've all gone through the process of purchasing airline tickets, taking into consideration price, frequent flyer miles, schedule, and other factors to arrive at our decision.

What if we could understand this behavior and use it to make predictions about what people will purchase and at what price? Extensive research literature exists on understanding consumer choice, and quantitative marketing professionals in all industries devote considerable energy to finding such information in large databases. Until recently, much of the activity involved studies ranging from several months to years, and they might have focused on long-term issues like which option packages to put on a car or what flight schedules to offer during different seasons. With more powerful computers and the proliferation of real-time data, however, consumer choice models are slowly gaining attention in operational pricing circles. Frequently updated models at a very granular level— for example, by region as opposed to nationwide—are under examination. Where these efforts will lead remains an open question. Consumer choice is, after all, an illusory concept. It's not clear that any particular individual understands the actual process she goes through to make a decision, much less that mathematical models can adequately predict what she will choose.

The questions surrounding unconstraining, buy-down, competitor price information, and consumer choice all help to demonstrate one of the great "secrets" of scientific pricing: it's not an exact science. Pricing scientists are continuously investigating new ways to improve their models of the world as business conditions change. Natural disagreements arise as pricing scientists argue about the validity of a new data source or the importance of a particular explanatory variable. Papers are published, conferences are held, and companies evaluate what does and doesn't work in their environment. Pricing science is a science, not an art, but there's art to the science of pricing.

As much as we want pricing science to be as exact as the laws of celestial motion, it's not. However, it's important to constantly remind ourselves that understanding everything that affects what people will pay isn't the goal of scientific pricing. The goal is to understand enough to make *better* pricing decisions. Omniscience might lead to huge revenue increases, but most companies know that just a few percentage points will make an important difference.

Requiring too much from pricing science is asking for disappointment. Even as pricing scientists develop better and better models, there won't come a day when everything is priced perfectly. Inherent statistical fluctuations can never be completely accounted for no matter how sophisticated our models become. However, this shouldn't stop us from playing the best game we can. We can't expect to win everyday, but using science to make rational decisions day in and day out ensures our eventual success just as surely as doing nothing ensures our failure.

* * *

The final hand of the 2005 World Poker Tour Championship came with both players holding an almost identical number of chips. Paul Maxfield found himself all-in with a king of spades and a five of diamonds, while Tuan Le held the king and jack of diamonds. Before knowing Maxfield's cards, Le had a 61 percent chance of winning. With Maxfield's cards revealed, the probability immediately jumped to 70 percent. Pairing the king wouldn't do Maxfield any good, since Le also held one and his jack was higher than Maxfield's five.

The three-card flop brought the jack of hearts, ten of hearts, and three of spades. With Le pairing his jack, Maxfield could no longer pair his five and win. Le's chances of winning jumped to a staggering 96 percent, and Maxfield needed a miracle.

To the surprise of everyone in attendance, the next card brought part of that miracle in the form of the queen of hearts. With a ten, jack, and queen on the board, and with both players holding a king, either an ace or a nine would make a straight. While Le couldn't lose, he saw his chance of winning plummet to 82 percent. If both players made a straight, the pot would be split and the game would continue.

After waiting a sufficiently long time to build the suspense, the dealer tapped the table, discarded, and turned over what would prove to be the final card of the game: the seven of hearts. With a slight chip lead over his opponent when the hand began, Tuan Le was the winner.

Both Le and Maxfield had their share of good fortune during the tournament, but no one makes the final table on luck alone. Winning requires a scientific knowledge of what's likely to occur, then using that knowledge to create an advantage. Maxfield hadn't played well enough to take the title, but he had beaten 450 other players, and for his efforts, took home $1.7 million, demonstrating that it isn't necessary to play perfectly to be a big winner.

After seven stress-filled days, Le raised his hands to the sky. Running to the audience, he located friends and family members, hugging, shaking hands, and giving high fives. A courageous player, Le had outwitted his opponent to become the 2005 World Poker Tour No Limit Hold 'em champion and pocket

$3 million. He'd known the odds. He'd made the right moves. He'd put himself in an advantageous position. And, in the end, he'd won.

But then after all, what did you expect?

Notes

- Unconstraining demand observations is not unique to the airline industry, with work in the business and economics literature dating back to Tobin (1958). Zeni (2001) discusses many of the more popular techniques used for airline revenue management.
- The buy-down phenomenon has long been recognized in revenue management settings, but has only received explicit attention in recent years; see Boyd, Kambour, Koushik, and Tama (2001) and Cooper, Homem-de-Mello, and Kleywegt (2006).
- Consumer choice models, as they are discussed here, are treated in the literature under the name *discrete* choice models. Many good books can be found on the topic, Train (2003) and Ben-Akiva (1985) among them. A general framework for revenue management incorporating discrete choice models can be found in Talluri and van Ryzin (2004).

CHAPTER 7

Upon Arrival: Hotels, Rental Cars, Cruise Ships, and More

People in America like to travel. With annual expenditures of $1.3 trillion on travel and tourism, $1 out of every $10 dollars accounted for in the gross domestic product goes toward getting somewhere, staying somewhere, and doing business or having fun while there. We take planes, trains, cars, buses, and boats. We use hotel rooms at an average of two and a half million per day. We shop, visit historical sites, play at theme parks, gamble in casinos, eat in restaurants, engage in athletics, and treat our bodies to rejuvenating spa therapies. Along the way, we generate over $100 billion in tax revenue.

As we know from experience, there are many similarities between booking a plane flight and reserving a hotel room. We call or go online in search of price and availability, and if we find what we like, we make a reservation. We pay, or at least provide a credit card number, and promise to show up at an appointed time and place. If our plans change, we reschedule or cancel, and as a result receive a credit or refund. It should come as no surprise, then, that hotels—along with rental car companies, cruise lines, and other travel providers—use operational pricing techniques similar to those used by airlines.

There are many reasons travel providers have been quick to adopt airline pricing practices, but three stand out in particular: similarity of distribution networks, similarity of business problems, and similarity of the underlying mathematics.

Similarity of Distribution Networks

When we travel, we don't just book flights. We book everything necessary to make our trip successful and enjoyable. Travel agents and Internet travel sites supply the many different components of our travel needs. Some Internet sites, such as Expedia, Orbitz, and Travelocity, provide one-stop shopping in the same way as travel agents. Others focus only on their own products. Hertz rents cars and nothing else. In the background, each travel provider maintains the equivalent of a reservations system to know who's coming and when, and so they don't sell too much inventory.

Figure 7.1 shows how a typical travel provider sells products in the market. Because travel providers sell to the same customers, it was only natural for global distribution systems—descendants of the early reservations systems—to begin handling more than airline tickets. Travel agents and global distribution systems were once the largest and most powerful distribution channel in the travel industry, and they remain important to this day. However, their dominance has been challenged with the arrival of the Internet. With the first online ticket sales occurring in the mid-1990s, approximately 42 percent of all sales transactions in the travel industry took place over the Internet in 2005. By 2007, that number is expected to rise to 55 percent. While many of these transactions took place through a Web interface connected to a global distribution system in the background, the total number of bookings through global distribution systems continues to shrink as a result of changes in technology.

The historical dominance of global distribution systems was first and foremost about computer hardware. Before the Internet, electronic communication wasn't as easy as getting an email account or setting up a Web page. The companies behind global distribution systems made huge investments in mainframe hardware to handle transactions and networking hardware to facilitate communication. The cost associated with building an electronic distribution network was extremely high and limited the number of entrants to the travel distribution market.

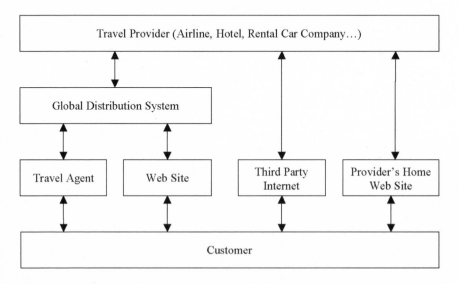

Figure 7.1 Four important distribution channels for travel providers.

With the arrival of the Internet, selling travel products became almost as easy as deciding to do so, and as a result, third party Web sites operating under many different sales models began appearing. From a business perspective, the tough part of setting up a new Web site was getting travel providers to let you sell their inventory. This difficulty was mitigated by the fact that newcomers were willing to charge substantially less than the traditional distribution channels for the right to act as middlemen. If a site was less expensive than established alternatives, there was relatively little cost to a travel provider to give the newcomer a chance. It also forced the traditional channels to rethink how they priced. Travel providers even went so far as to participate in the Internet bubble, either by demanding a share of the newcomers' companies in exchange for the right to sell the travel provider's products, or, in the case of Orbitz, by starting a new company owned by a consortium of travel providers.

In the late 1990s, third party travel sites began to appear almost weekly. Executives in charge of these sites spent venture capital money lavishly as they sought to speak with anyone willing to listen. During a visit to PROS by an Internet startup wanting to discuss a potential business partnership (translation: provide an entrée to PROS's client base), the executives made a pitch about why they, of all the Internet startups, were going to "emerge as the dominant player in the space." MBAs in hand and looking to cash in on one of the most bizarre periods in the history of American business, they were polished and practiced. Their sales model was predicated on consumers' willingness to engage in auctions for airline tickets, and an important part of their value proposition was how much fun customers would have.

Fun came up time and again during the discussions. I was willing to grant them that perhaps there existed a segment of customers that so enjoyed the thrill of bidding for airline tickets that they would visit the site for a periodic dose of adrenalin. It was even a nice change from the commonly repeated sales narrative of how the company would offer tickets more cheaply than anyone else. Still, it seemed to me that the big attraction of auctions was the potential for customers to get lower prices, and it wasn't at all clear that airlines would be willing to part with tickets if that occurred.

The most creative moment came as the executives described their plan to build market share. After presenting a list of the many companies they were in "serious discussions" with, the venture capital at their disposal, and their plans to get more venture capital, we were told in strict confidence that they were "in serious discussions with Leonard Nimoy" to be their spokesperson.

True, Priceline had signed William Shatner and had used him very effectively in its promotional campaign. But the thought of having the two leading characters from the 1970s cast of *Star Trek* vying for the public's travel dollars seemed absurdly funny. Of course, when Leonard Nimoy later did a short stint pitching for Priceline, I realized my calling wasn't in the softer side of marketing.

In addition to the traditional distribution channels and the new third party Internet channels that have gone on to establish themselves as viable businesses, home Web sites are an important means of distribution. With no one between the provider and the consumer, home Web sites cost less than other channels. But they have another important advantage: they allow the travel provider to communicate directly with the customer. This is true to an even greater degree with call centers, though call centers are much more expensive to operate.

Communicating directly with the customer has important advantages. For one, the travel provider is able to manage the customer's experience and thus the brand image. The Web sites of upscale hotels have grown very elaborate, with pictures and music designed not only to make you want to stay the night, but to spend $400 for the privilege. Communicating directly with the customer, companies can create an image of themselves serving business travelers, catering to families, or as heroes of the common man. The simple, straightforward layout of Southwest Airlines's Web site, complete with pictures of a happy workforce in their casual dress, is a marvel of branding.

Another advantage of communicating directly with the customer is the ability to control the actual sales process. A hotel selling its rooms through Expedia is only able to provide limited information about its facility, information that is presented in the same form as all the other hotels that are listed. Companies that want to sell their products in the best possible light can't do so when they hand this responsibility over to someone who is selling many competing products. The problems inherent in the relationship between providers and distributors aren't new. What's new is the ease with which travel providers can become their own distributors. Due to the many advantages of home Web sites, in the near future, they will continue to experience the fastest growth of all distribution channels at the expense of other alternatives.

Similarity of Business Problems

A second reason travel providers have adopted airline-style pricing is that the business problems are similar. Historically, traditional revenue management has been associated with the following business characteristics:

- *Non-negotiated pricing.* A price is posted, and buyers take it or leave it.
- *Fixed Capacity.* Hotels have a fixed number of rooms, just as rental car companies have a fixed number of cars, cruise ships a fixed number of staterooms, and airplanes a fixed number of seats.
- *Low variable costs relative to fixed costs.* Once a hotel or ship is built and staffed, the cost of filling an additional room is minimal. Car miles, on average, represent a minimal cost compared to holding and depreciation costs.
- *Inventory can be changed from one product into another.* Different products can be formed when hotel rooms, cars, and staterooms are essentially the same, just as airlines form different fare-class products from the same coach seats. Inventory hierarchies typically exist as well, in which products can be substituted one way but not the other. Airlines have larger seats in firstclass

cabins than in coach; suites are larger than standard rooms; cars come in different sizes; upper decks are preferred to lower decks. Hierarchies are dealt with through upgrading, providing a better product than the product that was sold.

- *Flexible return policy.* Since travel related transactions typically involve an agreement and not the delivery of a physical good, returns (cancellations) are easy. This necessitates overselling (overbooking) to compensate.
- *Segmentable market demand.* Segmentation is necessary to sell at different prices. Traditional revenue management models typically seek to design products that correspond to the desires of different customer segments, for example, distinguishing between "ocean view" and "garden view" rooms and charging appropriately. But market segmentation can also be achieved by charging different prices at different times.
- *Inventory is perishable.* Just as an empty airline seat loses its ability to generate revenue the moment the cabin door closes, so too do hotel rooms, rental cars, and staterooms lose their value when a day passes that they aren't used.

Traditional revenue management applications are often associated with computer-enabled sales, primarily because revenue management has historically been tied to reservations systems. It's difficult to practice revenue management in anything more than a rudimentary form without the aid of computers, but conceptually, computers aren't a requirement.

At times, the characteristics just listed have been used in an effort to distinguish revenue management from other forms of operational pricing. It's unlikely, however, that a precise definition of revenue management and a precise definition of pricing will ever be found. If anything, the distinction between revenue management and pricing is becoming increasingly blurred. It's a useful exercise to look at an activity that doesn't fall into the travel domain, for example, the sale of a garment in a department store, to see that traditional pricing problems aren't easily distinguished from revenue management problems.

- *Non-negotiated pricing?* Garments in a department store bear a fixed price.
- *Fixed Capacity?* Because there is a significant lead time between when orders are placed and when inventory is received, garment inventory often can be considered fixed.
- *Low variable costs relative to fixed costs?* The marginal cost of selling a garment is very low compared to the cost of operating the store and the cost of the garment itself.
- *Inventory can be changed from one product into another?* The single most significant departure from traditional revenue management, a peach blouse cannot be converted into a pair of tan slacks.
- *Flexible return policy?* Although returns occur, they aren't as frequent as in most travel settings. As a practical matter, this simplifies the problem.
- *Segmentable market demand?* Segmentation is achieved by varying price throughout the sales cycle, providing those who are anxious to buy an

opportunity to do so at a high price and then lowering the price as time passes. Ultimately, the garment may be moved to a bargain outlet.

- *Inventory is perishable?* A garment's value doesn't perish on a given date, but it loses significant value as the season wears on.

Similarity of Mathematics

A third reason travel providers have been quick to adopt airline pricing practices has to do with the underlying mathematics. Stepping back and abstracting as pricing scientists are known to do, the problems faced by travel providers are so closely related that at a mathematical level, they're almost identical. It's therefore common to hear scientists use such phrases as "a hotel is just a plane without wings," or "a rental car is just a seat with wheels." Of course these are oversimplifications, and each industry, much less each company within each industry, faces its own peculiarities. Still, as with all phrases that become part of a lexicon, there's a kernel of truth. Perhaps the greatest testimonial to this fact is when they are regarded seriously enough that people take issue with them. A recent journal article carried the attention-grabbing title "A cruise ship is *not* a floating hotel."

Taking hotels as an example, the mathematical similarities with airline revenue management are not difficult to see. Rooms in a hotel correspond to seats in a particular cabin, such as coach. Different prices are offered for these rooms by creating different products. The AAA rate for members of the American Automobile Association is one example of a product. Alternatively, rooms may be "deluxe" with special amenities such as bathrobes and turn-down service, or may be priced with a special breakfast, spa, or golf package. There may also be special rates for booking through a particular distribution channel, such as the hotel's home Web site. Hotels are fortunate in that, unlike airlines, the products they create represent more than price differences in the minds of consumers.

Just as airlines use fare class availability within a cabin, hotels use product availability to control what's for sale on any given day. With 100 rooms, "special rate" rooms may be limited to 50. Alternatively, opportunity costs can be calculated for

Table 7.1 The special room rate is unavailable for a single night stay on Wednesday since the price of $100 is less than the hurdle rate of $120. The special room rate *is* available for the 5 night stay from Sunday through Thursday, since the payment of 5 × $100 = $500 exceeds the sum of the hurdle rates: $490 = $80 + $90 + $110 + $120 + $90.

Day	Sun	Mon	Tue	Wed	Thu	Fri	Sat
Hurdle Rate	80	90	110	120	90	60	70

each room type on each day. A room rate is then available only if the total revenue it generates meets or exceeds the sum of the opportunity costs. In the context of hotel revenue management, the opportunity costs are frequently referred to as *hurdle rates*. Table 7.1 provides an example of how hurdle rates can be used to control sales.

Hurdle rates and other methods of controlling the sale of inventory help to alleviate *checkerboarding*. Checkerboarding occurs when short duration rentals block rentals of longer duration or, more exactly, when guest stays don't line up well. The example in Table 7.2 shows just how costly choosing the wrong customers can be, as rooms go empty even when demand for these rooms exists. By picking and choosing guests so that their stays fit neatly together, a hotel can improve revenues even without any variation in the nightly price of a room.

As with airlines, hotel reservations systems don't always provide for hurdle rate control, but some type of length-of-stay control is common. With min/max control, a hotel can set limits on the minimum or maximum length of stay for which reservations can be made. One of the more interesting aspects of hotel inventory control is that a guest may find that a hotel is unavailable on Monday night, but is available for a stay from Monday through Wednesday. Nothing magical has occurred to make a room suddenly appear. The hotel simply isn't willing to part with a room for one night.

The rental car industry deals with length-of-keep just as hotels deal with length-of-stay. There isn't an issue for cruise ships when everyone boards at the same time. However, if there are ports of call along the ship's itinerary where the passengers can embark and disembark, length-of-cruise becomes an issue.

Airlines also deal with the mathematical equivalent of length-of-stay—the number of flight legs a passenger uses. As discussed in Chapter 5, every time a seat on one or more flight legs is sold, it directly impacts any other itineraries

Table 7.2 Choosing customers 4 and 5 leads to revenue of $700 for the week. If guest 1 is chosen, guests 2, 3, and 4 are blocked from staying and the maximum revenue the hotel can generate is $300.

Day	Sun	Mon	Tue	Wed	Thu	Fri	Sat
Potential Guest 1				■			
Potential Guest 2				■	■		
Potential Guest 3		■	■				
Potential Guest 4			■	■	■	■	■
Potential Guest 5	■	■					

Special Room Rate = $100 per night

Hotel has one room

that would have used those flight legs, and indirectly affects the entire flight network. It's interesting to realize that, just like hotels, airlines could offer a single fare class with a single price on each flight itinerary, then increase revenues by intelligently picking and choosing who gets to fly. Of course, coupled with operational price changes by opening and closing fare classes, even greater revenues can be generated.

The picking and choosing problem has led to some noteworthy improvements in reservations systems. Travel agents book plane tickets day in and day out, and when they discover a loophole they know how to use it. Seeking to book a business passenger on flight 111 from Little Rock to Los Angeles, a travel agent might find the flight entirely sold out. However, there still might be availability on the flight itinerary from Little Rock to Hong Kong that connects through Los Angeles using flight 111 as the initial flight leg. If the agent tickets the entire flight itinerary, then cancels the Los Angeles to Hong Kong flight leg, the business passenger miraculously finds herself booked on the desired flight 111 to Los Angeles. To combat this practice, reservations systems providers introduced special functionality with the appropriately descriptive name *married segment logic*.

Table 7.3 shows how concepts in one industry correspond to others when the revenue management problems are translated into mathematical models. Looking at these conceptual similarities, it's not difficult to think of ways the list might

Table 7.3 How concepts in one industry translate into similar mathematical concepts in other industries.

Industry			
Airline	Hotel	Rental Car	Cruise
Seat	Room	Car	Stateroom
Cabin	Room Type	Car Class	Stateroom Type
Fare Class	Room Rate Category	Fare Tiers	Stateroom Rate Category
Flight Leg	Single Day	Single Day	Cruise Leg
Cabin Upgrade	Room Type Upgrade	Car Class Upgrade	Stateroom Type Upgrade
Flight Itinerary	Length-of-Stay	Length-of-Keep	Embarkation and Disembarkation Ports
Flight Network	Hotel and Specified Time Period	Rental Location and Specified Time Period	Ship and Ports

be expanded, and it hasn't taken long to do so. For example, pro shops at golf courses book up to four people in fixed starting times of roughly eight minutes duration. At a mathematical level, the starting times are identical to a four seat aircraft. Demand varies by day of week and time of day, and it's possible to displace foursomes by filling start times with twosomes or threesomes. Pricing and revenue management are now being adopted at different levels of sophistication by the golf industry.

Athletic events are another possibility. Sports teams have long used a rudimentary form of revenue management by charging different prices for tickets in different locations in a stadium. But thoughts are now turning to individual, regular season games. Games against teams competing for a regional championship are better attended than games against losing teams. Star players on a visiting team can attract a crowd even if the home team isn't doing so well. It's unlikely that we'll see ticket prices vary from game to game any time in the near future, but special discounts and promotions are even now being managed more effectively to take advantage of variations in demand.

Restaurants represent yet another area in which pricing science has seen activity. A restaurant has a fixed number of tables, and there are many ways to manage those tables for increased revenue. "Happy hours" are a basic but effective effort to bring people into a restaurant early when demand is low, and to move price-sensitive customers away from peak periods when space is at a premium. Happy hours also generate demand that might not otherwise materialize without lower prices.

Steering call-ahead reservations and walk-up traffic are other, more involved ways of managing a restaurant's inventory of tables. With each table having a certain number of seats, it seems natural to think of them as "planes without wings." But here, the underlying mathematics varies considerably, primarily because the length of time diners stay is unknown. Also, it's often possible to push tables together or pull up extra chairs. Planes with a flexible number of seats exist, but aren't common in many parts of the world. Some restaurant research has avoided the question of managing demand altogether, instead focusing on determining the number of tables, their size, and their configuration to better fit expected demand. If a restaurant is seating too many parties of two at tables for four, there's money being left on the table—so to speak.

Symphonies, ballets, plays, concerts, apartment complexes, storage facilities, charter agencies, equipment rental; the list goes on and on. The most provocative paper to appear in the research literature assesses the applicability of pricing and revenue management to the thriving, and legal, sex trade in Edinburgh, Scotland, begging the question of whether pricing may actually be the world's oldest profession. No matter what the problem, pricing scientists are taking their experience and their mathematical models with them as they conquer new and uncharted domains.

* * *

Visiting Walt Disney World in Orlando is always a treat, if a bit disorienting. The grounds are diligently cared for. Cast members—and everyone from dishwashers to the resort CEO are referred to as such—begin the day early as they scrub, sweep, rake, and water everything in sight. There are no signs of peeling paint nor grime to be found anywhere. Busses run on time. People smile. The top tier hotels are grand and playful—family destinations unto themselves. While it would be difficult to confuse a stay at the Animal Kingdom Lodge with being lost in Africa, or a stay at the Wilderness Lodge with a visit to a national park, they're great experiences nonetheless. For those who like a hot shower, a good meal, and endless entertainment for the kids, they're better than the real thing.

Disney, as anyone who works there will tell you, is a *magical experience*. It's a term used to describe what guests (never "customers") can expect during their visit. And it's something the organization doesn't take lightly. Management knows that the most important asset Disney has is its market image.

Riding the plane on my most recent trip to Orlando, I sat next to Casey, a three-year-old who was traveling with her brother and parents. Charming in a way that only a child her age can be, she lived in a world free from the complications that clutter our lives as we grow older. Was she going to visit Mickey, I asked? Yes, she replied. But then, with a little prompting from her mother, she told me who she really wanted to see. Her face radiated as she said, "I want to see *Cinderella*." And why not? Cinderella—honest, kind, and steadfast, facing unfair treatment with fortitude—it would be difficult to choose a better role model. And even if Casey later discovers that life doesn't always provide us with a fairy godmother, a handsome prince, and a happy ending, why shouldn't she have them now? I was sure Casey would meet her heroine, dressed in blue with long white gloves and yellow-blond hair, who would hold Casey's hand and make her feel like she, too, was a princess. Casey would never forget her visit to the magical kingdom that is Disney, nor would her parents, nor would the countless other children who disembarked the plane with me that day.

While Disney has always recognized that it's first and foremost a business, the corporate culture at Walt Disney World is nothing like that of a trading firm on Wall Street. Yes, money matters, and yes, it matters a lot. But everything emanates from the resort's ability to provide a magical experience—through its theme parks, its accommodations, and its commitment to service and entertainment. Of course, when it comes to hiring the right people, the resort has means of enticement that other businesses don't share. After all, who wouldn't want to work at Disney?

Mark Shafer is one of those who heard the siren's call. A former director of international revenue management at Continental Airlines, Shafer joined Walt Disney World in 1995 as manager of resort pricing and revenue management. A man of medium build and military shoulders, he speaks in a powerful baritone that allows him to successfully use the word *awesome* where others would fail. "You've gotta see the new ride at the Epcot," I was told during one visit. "It's *awesome*." And later that afternoon, I found myself seated in *Mission: Space*, experiencing forces my body had never felt before.

Awesome isn't reserved for theme parks. Shafer's greatest joy is pricing and revenue management, and on more than one occasion, he's used *awesome* to describe mathematics. If the pricing community was ever in need of a poster child, Shafer would win hands down. When he's not scribbling pictures and formulas on a whiteboard, an activity that occupies 25 percent of his time but could easily occupy all of it, he's preaching the Good Word from the Book of Revenue Management to anyone who will listen. He's been chastised for reading mathematical texts on vacation, and he keeps a stack of books on his shelf to distribute to new employees or anyone in the Disney organization who needs to learn a little bit more. To Shafer, there's a world to be revenue managed, and he seems intent on taking personal responsibility.

When he arrived in Orlando, revenue management wasn't an organizational priority. But rather than view this as a limitation, Shafer viewed it as an opportunity. His first order of business was to get a good revenue management system in place to manage the 26 thousand rooms located on the property. Competing with many other initiatives, Shafer and his colleagues were able to secure funding for a system and contracted with PROS.

No project goes flawlessly, but when the revenue management system was up and running, it was considered one of the most successful implementations in recent years. With the project behind him, Shafer was able to turn his attention to the bigger Disney picture. Promoted to director of resort pricing and revenue management, and later to vice president of revenue and profit management, he began a campaign to broaden the awareness of the power of revenue management within the organization. Early on, at Shafer's request, PROS scientists showed up with PowerPoint presentations on dynamic programming and Bayesian hierarchical forecasting, giving lectures to managers, directors, and vice presidents from different parts of Disney. While the presentations made an impact, they left a large hole between science and exactly what that science meant for each person in the room. Shafer found himself faced with the perennial problem facing all science-based business projects: how to make the science real so that people would adopt it.

A look at what Shafer uses today shows exactly how far he's come on the educational front. Seated in his office, I was presented with a pair of CDs prepared by one of his managers. Stamped with the departmental logo—a tilted pair of mouse ears outlined with an upward pointing arrow—one read, "Restaurant Revenue Management Library" and the other, "Retail Revenue Management Library." The disks contained hundreds of articles, reports, papers, and presentations representing the two areas in which Shafer had most recently focused his attention. "I like to use these to show that we're not mad scientists, that we're not out on the bleeding edge doing this stuff all by ourselves," said Shafer.

Feeling that it helps add credibility to his efforts, Shafer likes to bring in acknowledged experts from outside the company, many from academic institutions. He also encourages people to attend conferences on pricing and revenue management, recognizing that there's nothing quite as powerful as being immersed in the baptismal fountain. A list of Disney attendees at a recent conference showed a surprising level of participation, most from departments that don't report to Shafer. I asked how he accomplished this. "I paid for them to go from my own budget," he said.

Reaching onto his shelf, Shafer picked up a carefully tabbed three ring binder, one of many lining his wall. "Let me show you what we've been doing to make a case for revenue management at our restaurants," he said.

One of the big issues in revenue management for restaurants is turn time, the time it takes a table to be served and cleaned in preparation for the next party. Faster turn times mean more guests will get served in the course of an evening, generating more revenue. Disney periodically sends teams of industrial engineers to measure average turn time at its restaurants. This information is then used to determine how many reservations each restaurant can take and how widely spaced they need to be. But Shafer and his team realized the importance of something that had long been ignored at restaurants on the Disney property. "Not everyone takes the same length of time," said Shafer. "It's the variability that poses both problems and opportunities."

Table 7.4 Components of a slide describing how science can be used to reduce wasted table time and guest waiting time.

	Turn 1	Turn 2
Table 1	60	60
Table 2	60	60
Table 3	60	60

On average, tables are occupied for 60 minutes

Occurs:

	Turn 1	Turn 2
Table 1	45	Next Party
Table 2	60	Next Party
Table 3	75	Next Party

When Table 1 is occupied for less than 60 minutes, it sits idle. When Table 3 is occupied for more than 60 minutes, the next party must wait.

Future Probabilistic Approach:

	Turn 1	Turn 2
Table 1	45	Next Party
Table 2	60	Next Party
Table 3	75	Next Party

Understanding variability alleviates wasted table time and the time guests must wait.

Table 7.5 Components of a typical dining experience. Much of the time is under the control of the restaurant.

Guests Seated	1 minute		
Greet / Order Drinks	3 minutes	Service Time	Owned by Restaurant
Deliver Drinks	1 minute		
Order Meal	16 minutes		
Deliver Meal	20 minutes	Dining Time	Owned by Guest
Order Dessert	5 minutes	Service Time	Owned by Restaurant
Deliver Dessert	6 minutes	Dining Time	Owned by Guest
Request Check	2 minutes		
Deliver Check	2 minutes		
Collect Check	1 minute	Service Time	Owned by Restaurant
Return Check	1 minute		
Guest Leaves	2 minutes		
Bus Table	5 minutes		

I was shown a printed copy of a PowerPoint slide that Shafer used to explain how revenue management applies in a restaurant setting (see Table 7.4). "If the turn time averages 60 minutes, we would separate our reservations by this much. But if someone gets up and leaves 15 minutes early, the table sits idle for this long. If someone stays an extra 15 minutes, then guests have to wait." With his slide and his brief explanation, the problem was crystal clear. I liked the fact that the final picture on the slide showed parties perfectly lining up, with no wasted table time and no guests waiting. By its very definition, variability implies that things can never go this perfectly. But Shafer wasn't selling the details, he was selling the concept, and at the conceptual level, he was absolutely correct. Accounting for variability will make things go better, and as a result, reduce turn time.

Shafer shared a second slide showing the basic components of a meal (see Table 7.5), a result of discussions with Dr. Sheryl Kimes of Cornell University. "We usually think of the length of time a guest dines as uncontrollable, but it's not. We control a lot of it," said Shafer. Delivering drinks and menus, taking orders, preparing checks, bussing tables—all of these activities contribute to turn time. And in each case, speeding things up enhances the customer's dining experience. "It's hard getting people to understand that revenue management isn't synonymous with raising prices. A lot of the time it's just about using your inventory better." And this refers not just to table inventory, but also to the inventory of servers. If additional staff can increase throughput and revenue by reducing turn time, then hiring new cast members makes sense.

Thinking holistically about the dining experience seems obvious, but it's not. Savings achieved by reducing costs, like labor costs, are easy to calculate and execute. More nebulous is the increased revenue that can be achieved by reducing turn times.

Going to the whiteboard, Shafer discussed another idea he's working on. "We have 47 restaurants where guests sit down and are served at tables. Some of the restaurants are very crowded, while we have other great restaurants that are less likely to fill up because of location or because guests just aren't aware of them. Doesn't it make sense to have whoever handles dinner reservations suggest restaurants that are less crowded?" I agreed it makes a lot of sense. Throughput, Disney revenue, and customer satisfaction can all be increased—there aren't any losers. It simply requires getting the right information to the right people and training them what to do with it. But what about resort guests who really want a specific restaurant, or a particular price range, or a geographic location, I asked? The answer was obvious. "There's nothing that stops them. But it sure doesn't make sense to force someone to wait at a busy restaurant if they don't want to." Having 47 options, I realized, has its advantages.

Shafer's experience at Continental, working with the international side of the business, taught him many lessons. Business practices in Asian markets were so different from the practices surrounding domestic travel that revenue management required rethinking. Seats were sold in huge blocks to third parties who in turn resold them, as opposed to the domestic model in which the airline sold seats to customers, paying a fee to whoever was involved in the sale. Shafer knows from experience that it's necessary to get your hands dirty understanding a business problem before you start solving it. As he and his team undertake new projects, a commitment to dirty hands is mandatory—figuratively speaking.

Team members working on the restaurant project were required to work full shifts at different restaurant positions. Whether they were cooks, waiters, or receptionists, team members learned firsthand what was and wasn't possible, not just what looked good on paper. Restaurants throughout the resort were visited to better understand their similarities and differences. Just like the Disney Imagineers (Disney's term for their creative teams) immersed themselves in the African experience before building the Animal Kingdom Lodge, Shafer and his team spent weeks washing dishes, serving food, and bussing tables to better understand food service. The commitment to learning wouldn't go unnoticed by the restaurateurs, either. The team would never be food and beverage insiders, but a willingness to walk in food and beverage shoes would go a long way.

Waiting for Shafer to return to his office, I was passed by John, a manager in Shafer's group, holding two large Disney shopping bags. He was in search of coat hangers, unhappy that he had forgotten to ask the woman at the checkout counter to leave them with the clothes he'd purchased the night before. Having located what he was looking for, John removed two glittering dresses from the bags. Accompanying them in separate packaging were costume jewelry, sparkling shoes, elbow length gloves, a magic wand, and a wig. "Which one do you like best?" asked John about the dresses. Somehow, the question didn't seem the least bit strange. I told him I liked the blue one.

"Yeah, me too," said John. "That's the standard Cinderella. The other one's the wedding dress." I shook my head in agreement and wrote this in my notes, realizing later that we were both experiencing a memory lapse about the story. Cinderella went to a *ball*, and she'd worn a *ball gown*. The glass slippers should have made this fact obvious. It was the beautiful blue peasant dress, however, that threw us off. It should have been soiled, the result of sweeping the floors and cooking the food and all the other activities demanded by the wicked stepmother. But this would have defeated the purpose of making it attractive to buyers.

John had been on a scouting trip for Shafer's other ongoing project, merchandise sales. The team was looking at sales figures for the various items now on display, and they were trying to understand why some of the items were selling so much better than others. Evaluating the individual items was important. We all agreed that the wig, sitting by itself in the bottom of a plastic hatbox, was a little creepy—a victim of the recent generation of horror films. But evaluating the items as a group was important, too. How did they look when placed next to each other? When the jewelry was placed on the dresses? When the glass slippers were hung rather than left on the floor?

Back in Shafer's office, he pulled down yet another of the three ring binders. Inside were detailed color pictures describing various aspects of the guest's shopping experience. Line of sight was one topic. As guests walk through stores, certain areas are blocked from view while others show prominently. Shafer wasn't just interested in static views, but in the dynamic views guests encounter as they walk through a store. Do they enter an aisle from the left or the right? An elegant picture depicted paths that guests might traverse as they made their way through one particular store. There was a picture showing the importance of height in merchandise placement. Was it important to place an item at eye level? There were pages depicting different categories of merchandise and how sales are impacted by traffic flow patterns.

Nothing Shafer showed me was new. Retailers have developed detailed rules of thumb over the years, some from experience, others from studies. Beverage distributors want their product to be displayed at the end of the aisle or stacked on the floor. Eye level shelf space is coveted, and retailers are savvy enough to understand that the definition of eye level changes with who they are selling to. How Shafer's approach differed from what had been done in the past was in the focus on letting data drive decisions—constantly, and on a store-by-store basis. The basic retail principles are known, but there are so many variables that it's necessary to dig through the data to find out what works and what doesn't, and to check data that doesn't make sense by looking at the actual merchandise.

Cinderella's dresses were one example, but Shafer described another instance in which data led his team to visit one of the stores. A football hadn't been selling well at $19.99, so the price was reduced $9.99. Yet data showed that sales hadn't changed at all. To Shafer and his team, this didn't make sense. Price wasn't everything, but it was difficult to imagine that shoppers wouldn't respond at all to such a drastic price cut.

Arriving at the store, the answer was immediately apparent. New price tags had been put on the boxes containing the footballs, but they were located on the bottom, completely out of sight. There was nothing to indicate the discounted price to passing guests unless they picked up the box, turned it over, and looked at it. The price reduction wasn't driving any new sales because shoppers didn't know about it.

Like Shafer, Neil Corbett appreciates the power of data. Director of revenue and profit management, he works closely with Shafer as they take on ever more responsibilities. Corbett has experienced every aspect of running a hotel during his 25 year professional career in the hospitality industry, from the front desk to general manager.

Entering Corbett's office, I immediately noticed an entire wall lined with pictures of Disney villains—Cruella DeVil, Jafar, Ursula, Scar, Captain Hook—it only took a second to realize that great stories need great villains as much as they need heroes. "That's who we are in revenue and profit management," said Corbett, "whenever sales calls."

Corbett was referring to the Disney sales agents who have the responsibility for booking large groups and conventions. As in most organizations, their commissions are based on volume. It's Corbett's responsibility to make certain they book business that makes sense. Repeating a common theme, Corbett recounted how sales agents like to spin things when they call him: the conference organizer needs a price break; the event must absolutely take place the week of March 20; it must be held at the Grand Floridian. "I know all the stories," said Corbett, "I used to *be* a sales agent."

Attrition clauses, in which group organizers guarantee to pay for a minimum number of rooms no matter how many people show up, are a constant source of frustration. Sales agents write attrition clauses into contracts then routinely call to have them forgiven, citing the undeniable fact that no one in the industry really enforces these things. "I don't like to be the enforcer," said Corbett. "I'd rather have guests in the rooms enjoying the theme parks than try to collect charges for empty rooms. But there's a huge opportunity cost when we commit inventory to a group and they don't show up." His team also has to play tough with those who want to book short stays during popular holiday periods. "New Year's Eve is a great example," he said. "Lots of people want to book a room for New Year's Eve, but when they all leave the next morning a hotel can be empty for days. I want to fill the rooms with families looking to stay for the week."

Most recently, Corbett's been focusing on transaction data, or "transactions within a transaction," he told me. Recognizing that the magical experience starts the moment someone calls to inquire about a stay, Disney handles the majority of its bookings directly, through a call center or over the Internet. In this respect, Disney is unlike many companies in the travel industry that sell their products through the elaborate travel distribution network. The advantage is that Disney has much closer contact with the people they're selling to. If you came last year and had a wonderful stay at the Boardwalk, telephone sales agents can use that information as a starting point for helping you book this year. As a call proceeds, there are many different ways it can unfold. Understanding what works best in different situations can lead to increased conversion rates—turning shoppers into buyers.

An example of managing a transaction is determining what to do when a shopper's preferred resort is booked. Should the agent recommend a different date? A different resort? If a different resort, which one? These are complicated questions. But with data, a company can begin to understand at a detailed level what works and what doesn't.

"Some of our core ideas have been shaken," said Corbett about the recent effort analyzing transactions. He shared one such idea in particular. Disney operates a collection of value resorts that are all very similar in layout—All-Star Movies, All-Star Music, and All-Star Sports. A visit to any of the properties makes it clear that the buildings, rooms, and common areas were built from the same floor plan. What differs is the theming—the colors, the art on the wall, the giant sculptures on the grounds. Where the All-Star Music has a guitar, the All-Star Movies might have a motion picture camera, and the All-Star Sports a baseball bat. The prices at the different resorts are the same, as are the amenities.

With so much similarity, initial thinking was that the properties were interchangeable: if shoppers called to make a reservation at one resort but it was unavailable, they'd be happy to take a room at either of the other two. Data about booking patterns showed this wasn't the case, demonstrating that even the best companies can fall prey to faulty preconceptions. If there was ever a company that understood the importance of theming and what it means in the minds of the public, that company is Disney. Yet here was a case where building layout had clouded the obvious. Movies, music, and sports are very, very different.

Shafer and Corbett are happy with how far they've come. Support now comes straight from the top of the organization, showing just how important scientific revenue initiatives are considered at Walt Disney World. There remains much to do. Disney has theme parks around the world, and Shafer's success has landed him with global responsibilities. The best practices from Orlando, Anaheim, Hong Kong, Tokyo, and Paris can now be leveraged worldwide. Arcades have yet to be touched. Disney owns assets that extend beyond the realm of the theme parks, including several media outlets. There's always new science to be tested and incorporated in the systems. Many people would consider the challenges overwhelming, but for the most part, Shafer looks like a kid in a candy shop—or perhaps a kid turned loose with a lifelong pass to Disney theme parks.

On my way out of the office, I stopped to take a look at pictures of Shafer's wooden boat. It wasn't big, a rowboat perhaps a dozen feet long, but he'd done all the work by hand over a period of years. There were pictures of how the work progressed board by board. Each needed careful shaping before it was bent and glued into place. Then there was sanding and countless coats of varnish. It seemed that the end might be in sight, though Shafer wouldn't commit to a date.

Along with the pictures were names proposed by members of his team. The list contained office standards—*Speedy, S.S. Minnow, Titanic*—but there were a few special names that caught my eye: *The Price, The Incrementality, The Displacement*. I asked Shafer his thoughts. "I'm leaning toward *EZ Displacement*," he said, pointing out that "E" and "Z" are the first initials of his sons' names. *EZ*

Displacement. It was right. But I had to wonder if the name would ever find its way onto the boat. After all, that would be an admission that the work was done, and all the fun along with it.

Notes

- Readers interested in learning more about applications of revenue management in the travel and transportation industry are directed to Yeoman and McMahon-Beattie (2004) and to the Palgrave Macmillan publication the *Journal of Revenue and Pricing Management.* See also Kimes and Schruben (2002) and Kimes (2005) for a further discussion of golf and restaurant revenue management.
- In addition to the individuals named in the text, John Quillinan provided valuable source material contained in this chapter.

CHAPTER 8

The Just Price

I t happened during one of our family trips to eastern Washington. Two parents, two kids, one dog, and a station wagon, we were the statistically correct 1960s family. My father didn't like to waste vacation time and preferred leaving the San Francisco Bay area in the afternoon and driving all night. It was late in the evening. My mother was dozing in the front seat; my younger sister was asleep in the back with the suitcases. I couldn't sleep, and as I sat stroking the head of our white German Shepherd, I was in a thoughtful mood. My father was going to pay for it.

"Dad, when you say we're two-thirds of the way there, is that more or less than a half?"

"More."

"How much does a speeding ticket cost?"

"Depends on how fast you're going."

"How much is something worth?"

"Exactly what someone else is willing to pay for it."

Here was something to think about. An item's value was related to its price. Or was it? I knew the family car was expensive, and we needed it to get around. I knew our house was expensive, too, and we had to live somewhere. These items fit with my father's claim that items of high value carry a high price. But there were also important things that cost almost nothing. I would have paid anything to protect my ragged, old blanket, but my mother had told me it had cost only a few dollars.

Price and worth are fundamentally different concepts, but they're deeply intertwined. In our daily use of the English language, we frequently treat them as siblings if not synonyms. "What's it worth to you?" is a question about our willingness-to-pay, not a request for a value judgment. When we ask, "What price, love?" it isn't normally in the context of negotiating payment for services. The confusion between price and worth stems from our ethical makeup, demanding that price somehow be connected with the inherent value of an item. We want price to reflect worth. We want a fair price. We want the *just* price.

Without question, airlines have impinged on our notions of the just price. When buying a new television, we look at the various models, compare features and prices, and purchase the one that best meets our objectives. We may be disappointed that the television we really want is out of our price range, but we rarely walk away angry at whoever set the price. On the other hand, buying an airline ticket can bring out a variety of emotions: exasperation, frustration, anger, and occasionally even joy. We've tempered our emotions as we've grown more used to the way airlines price, but we still feel them. There's something that's not quite right about airline pricing that leaves us disquieted and shaking our heads.

Why do we feel this way? What is it about airline ticket pricing that reaches deep into our psyche and leaves us feeling troubled? Today, we rarely express our trepidations about price in terms of justice, thanks in large part to Adam Smith's 1776 publication *The Wealth of Nations*. In Smith's world, there is nothing just or unjust about price. Price simply *is*; it is a natural phenomenon that comes about as a result of exchange. Smith didn't abandon the connection of price and worth, but the importance of worth was subjugated to prices established in the market. With Smith, a whole new emphasis was placed on understanding what the market *does* to determine prices rather than what it *should be doing*. In the spirit of the Enlightenment, economics was freed from traditional thinking and allowed to emerge as a respected empirical science.

Yet throughout most of history, the connection between price, worth, and justice occupied the attention of some of the world's greatest intellects. Ethical questions were foremost, but were always balanced with the practical considerations of running an orderly society. Governing bodies needed to determine if and when they should get involved in disputes over price. Philosophers, theologians, and legal scholars all contributed to the debate over the just price from antiquity through the Middle Ages, a time during which price was often held to be a moral issue as much as a market issue. The interaction of these groups in turn provided tremendous insight into our beliefs about justice in pricing before we abandoned them, at least in practice, to the free markets of Adam Smith.

* * *

Together with Socrates (470–399 BC) and Plato (427–347 BC), Aristotle (384–322 BC) is without question one of the most influential figures of all time. A respected thinker and teacher during his lifetime, Aristotle was a scientist and philosopher, a politician and psychologist. Tutor to Alexander the Great, Aristotle's writings often showed a profound worldliness about them. While Aristotle pondered questions of meaning and the existence of the soul, he was not afraid to tackle more down-to-earth questions as well.

Aristotle's treatment of rhetoric, which to the Greeks simply meant the art of persuasive argument, was a good example. In contrast to Aristotle, Plato denounced rhetoric, believing that logical arguments led to truth and that rhetoric only created confusion. Plato's position was in part a response to a society of teachers known as the sophists, whose practice of rhetoric had evolved to a point where the truth of an argument was irrelevant. But it was also attributable to a particular mindset—a mindset that failed to understand the importance of looking through other people's eyes.

Alternatively, Aristotle fashioned a respectable vision of rhetoric that dominated Western thinking through the Middle Ages and remains influential to this day. Like Plato, Aristotle firmly believed that logic was essential when constructing an argument. But he went further, emphasizing things like appealing to the emotion of listeners, good presentation skills, timing, and how the mood of an audience influences what they hear. Where Aristotle differentiated himself from the sophists was in his focus on the process of creating a persuasive argument rather than on winning at all costs. He brilliantly clarified his position in the very first sentence of his book *Rhetoric*, in which he refers to rhetoric as the counterpart to Plato's logic. Logic was required to *find* truth, but rhetoric was necessary to *communicate* truth. Aristotle understood what Plato didn't, and what many present day scientists have yet to learn.

It is from Aristotle that we discover some of the earliest thoughts on the relationship between price and worth. In Book 5 of the *Nicomachean Ethics*, Aristotle distinguishes between two types of justice. *Distributive justice* is related to questions of how honor or community wealth should be apportioned, while *corrective justice* is "that which plays a rectifying part in transactions between man and man." He further subdivided corrective justice into transactions that are "voluntary" and "involuntary," while providing examples of each. Voluntary transactions include "sale, purchase, loan for consumption, pledging, loan for use, depositing, [and] letting." Involuntary transactions include "theft, adultery, poisoning, procuring, enticement of slaves, assassination, false witness . . . assault, imprisonment, murder, robbery with violence, mutilation, abuse, [and] insult."

Of interest is the fact that corrective justice includes commercial transactions along with what, for the most part, are acts of crime. Aristotle did not intend to equate commerce with crime, but he did see many similarities. When something is stolen, a balance or state of equality is disturbed. The thief gains at the expense of the individual he's stolen from. Justice is served when the loss and the gain are equalized. Commerce is similar. Parties enter into transactions because both can gain by doing so, but "proportionate equality of goods" is essential for a just transaction.

With a focus on justice, Aristotle clearly had in mind some sense of worth or value when he spoke of equality. However, he also recognized that to examine just exchange in practical terms required determining when equality was achieved. His common standard of measurement was the "demand" of the individuals involved in a transaction, which scholar John Baldwin argues would be better translated as "want" or "need." Aristotle then connected "demand" to money and therefore price. Wrote Aristotle, "All goods must . . . be measured by some one thing. . . . Now this unit is in truth demand, which holds all things together . . . but money has become by convention a sort of representative of demand."

And he stated, even more strongly, that "all things that are exchanged must be somehow comparable. It is for this end that money has been introduced . . . for it measures all things."

For Aristotle, money is a convenient means of measuring the value of an item. But it isn't that we express how much we value something by the price we are willing to pay for it. Rather, each item has an inherent worth, and money simply facilitates the transaction. Aristotle was therefore left with the difficult task of determining the just price—the task of putting an actual number on the price tag.

Aristotle's description of how the just price is actually determined remains unclear and has been the subject of debate for centuries. However, as Baldwin points out, "in terms of everyday economic experience . . . [Aristotle's] justice of exchange was probably nothing more mysterious than the normal competitive price." Here we find the beginning of a question that philosophers and theologians would be faced with time and again. Even if we agree that there is a just price and that this price is related to the worth of an item, how do we determine it? To his credit, Aristotle was courageous enough to attempt an answer, something not all philosophers felt obliged to do.

Throughout history, much of the debate over the just price centered on mercantile activity—the purchase and resale of goods without actually producing anything. A trip to the local grocery store emphasizes the important role of merchants in our daily lives and the value they bring. Yet for much of history, profits from mercantile activity were viewed with contempt. Without producing anything, a merchant wasn't providing any value.

The economic importance of mercantile activity might have been recognized earlier if it weren't for the desire of many merchants to accumulate material possessions. Plato conceded that merchants served a useful function in society, but felt that the temptation to acquire excessive wealth was such a powerful force that their activities should be severely limited, even to the point of forbidding Greek citizens from engaging in the practice. Aristotle believed that merchants were acting appropriately as long as their profits are limited to cover their needs, but looked unfavorably upon the practice of amassing wealth for its own sake.

Unjust pricing and wealth are different concepts, but as the Greeks realized, when just prices were required to reflect underlying worth, it was a logical necessity that wealthy merchants were charging unjust prices. If the prices they charged were just, then, *de facto*, they couldn't be rich.

Wealth became the central theme as the Roman Catholic Church emerged to become the dominant voice on how people should behave. Christ's teachings carried a message of spirituality, a message that had little place for material goods. In the Sermon on the Mount, he asked God to simply "Give us this day our daily bread," and that "Your will be done on earth as it is in heaven." He traveled with his disciples, carrying few possessions, and depended on the charity of those he met to supply his basic needs. He did share the Greek concern over wealth, though in Christ's case, this meant straying from the spiritual life. When asked by a young man what he must do to earn salvation, Christ responded, "Go and sell your possessions and give to the poor." When the man left dejectedly, Christ reiterated the message to his disciples. "Again I tell you, it is easier for a camel to go through the eye of a needle than it is for a rich man to enter the Kingdom of God."

Christ did address price in one important way in the parable of the laborers in the vineyard. In this allegory, he tells of a landowner who hires workers at four different times throughout the day—the early morning, mid-morning, afternoon, and early evening. Although the workers spend different lengths of time in the field, upon reaching the end of the day, the landowner first pays those who were hired last, and everyone receives the same wage. Not surprisingly, this does not sit well with those who were hired first, and they complain about their payment.

> When they received it, they grumbled at the landowner, saying "These last men have worked only one hour, and you made them equal to us who have borne the burden and the scorching heat of the day." But he answered and said to one of them, "Friend, I am doing you no wrong; did you not agree with me for a denarius? Take what is yours and go, but I wish to give this last man the same as to you. (Matt. 20:11–14)

The intent of the parable is to communicate that the repentant can expect to share equally in God's Kingdom with those who live long and virtuous lives. This is yet another recurring theme in Christ's teaching. But in making his point, he chose an example that most people would consider a case of unjust pricing, and an egregious example at that: paying an unfair wage.

While Christ had relatively little to say on the just price, the early Church Fathers expounded upon Christ's position on the evils of wealth. Beyond the fact that wealth could divert humankind from more important spiritual goals, another emergent theme was that the very practice of commerce was wrought with ways in which humans could sin. Just as the Greeks mistrusted merchants, the Church saw mercantile activity as a swamp of immorality, complete with greed, lying, and cheating. Merchants weren't just tempted to stray from the will of God at every turn. A degree of moral corruptness was a necessary part of the job. As summarized by Baldwin, Augustine (354–430) maintained that "just as art cannot exist without imposture, neither can business exist without fraud." Tertullian (155–230) denounced the activities of merchants while missing entirely the important role of exchange in improving the human condition.

> Is trading fit for the service of God? Certainly, if greed is absent, which is the cause of acquisition. But if acquisition ceases, there will no longer be the necessity of trading. (Tertullian, De idolatria)

For the early Church, the point was not so much to understand justice in pricing as it was to encourage living in the image of Christ. The laws of exchange were left in the hands of the Roman Empire, which for centuries allowed prices to be determined in an open and largely unregulated environment. Roman law culminated in the systematic gathering and interpretation of legal tenets used throughout the Roman Empire under Emperor Justinian in the sixth century. Under the Justinian Code, it was understood that buyers and sellers would attempt to outwit one another in negotiations over price. Such actions were considered completely acceptable. The role of the courts was to ensure the smooth functioning of the social order in general, and commerce in particular. With courts refusing to intervene over issues related to price, there was no practical need for even thinking about the concept of a just price, much less determining one.

For this reason, an important exception to the *laissez faire* standard of Roman law known as the *laesio enormis* received considerable attention as theologians and legal scholars of the Middle Ages came to terms with one another. In its original form, the doctrine maintained that a man who was paid too little in the sale of land was entitled to restitution. The Justinian Code went on to define "too little" as an amount that was less than half of the true price. Of crucial importance was the suggestion that there exists a "true" price—some price that everyone can agree is about standard. In admitting the existence of a true price, even in one very specific instance related to the sale of land, the Justinian Code was not entirely ready to cast out the connection between price and justice. The law could get involved in disputes about price. This would prove important in later years as the Roman Empire crumbled and the Church became the dominant institution in Western society.

An important step in defining the Church's position toward mercantile activity came from Augustine in a document describing a (probably fictitious) conversation between himself and a merchant. After lambasting the merchant for damaging the good name of Christianity through his lying ways, the merchant successfully defends himself. Reminding Augustine of the Christian doctrine that "a laborer is worthy of his hire," the merchant points out that he is providing a service by transporting goods and therefore should be allowed a reasonable profit. Lying should not be confused with the profession. It was possible to live according to Christian ideals and be a merchant; it was just that many merchants didn't.

In accepting these arguments, Augustine appeared to be contradicting his position that business cannot exist without fraud. If anything, however, Augustine's position emphasized the difficulties faced by theologians attempting to reconcile Church doctrine with the very real commercial activities taking place all around them.

Augustine reached the same conclusion as Plato and Aristotle, that merchants provide value. Transportation, while not adding value to a good itself in the way a baker might do so by turning flour into bread, was nonetheless a service of value to the community. Bringing value of any type was a worthy undertaking. The problem was in distinguishing between the value brought by merchants and what we might now call excessive profits. If a merchant regulated profits by charging only an amount that helped maintain a reasonable living—cost plus an amount necessary for food and clothing—these profits were morally justified. Anything over and above this was a sin. Profits based on an effort to buy cheap and sell dear were among the most heinous. Buying a house and later selling it for more than the purchase price, having done nothing to improve the house in the interim, is an activity we engage in today without a second thought. To the early Church this represented the worst kind of mercantile activity.

From the eighth century through the end of the Middle Ages, just profits and the just price would receive extensive attention, as captured in Church documents from the period. Justice wasn't served simply because two parties agreed on a price or because both were happy with the exchange. While legal involvement normally arose only when one party was dissatisfied with a transaction, the Church used these opportunities to gradually develop a highly organized set of laws and theological standards from which it could pronounce judgment, both in the courts and in the eyes of God.

As late as the twelfth century, groups within the Church continued to decry the evils associated with being a merchant. But faced with the rapidly developing economy of this period, Church legal scholars agreed on a set of principles for dealing with mercantile exchange in practice. First and foremost, the free bargaining found in Roman law would serve as the accepted basis of determining a price, but within the boundaries established by *laesio enormis*. The question of whether the profits realized from the sale of goods were just and reasonable would be left to the individuals involved in the transaction and, ultimately, the confessional.

The Church nonetheless made it clear what was just even as it removed itself from in-depth involvement in commerce. A just profit could incorporate the cost of materials, the cost of labor, and even the cost of risk. Even further profit was acceptable as long as the intentions were noble, involving care for the basic needs of self and family or the support of charitable works, for example. While the doctrine of the Middle Ages was a far cry from the early Christian ideal of renouncing worldly goods and thus trade, Church legal scholars had arrived at a workable approach to issues of exchange without abandoning the concept of justice altogether. Or so they thought.

Even as Church legal scholars were formalizing a legal system in tune with the practical demands of the time, an active community of theologians was at work addressing questions pertaining to the just price from a purely moral position. Scholasticism, the rediscovery of the ancient Greek philosophers and the efforts to reconcile their teachings with those of the medieval Church, was at its high point. Starting with scattered interpretations and definitions, the theologians began by establishing a coherent, rational foundation for the study of justice.

The theologians and legal scholars disagreed on the concept of a just price. Whereas Church legal scholars accepted *laesio enormis* as a test for whether legal action could be taken, theologians held to the belief that any transaction in which a buyer paid more or less than the underlying worth of a good could not be considered just, even if that difference was a mere penny. It is not difficult to see how the theologians arrived at this conclusion, nor is it difficult to imagine that in concept the legal scholars were probably sympathetic. The disagreement was more than theoretical, however. The one-half recipe of *laesio enormis* was a long way from guaranteeing that price perfectly reflected worth.

Two Church intellectuals who took up the argument of the theologians were the distinguished scholastics Albert the Great (1193?–1280) and his student Thomas Aquinas (1225–1274). Exploring the nature of worth underlying the just price, the two men immersed themselves in the recently rediscovered *Nicomachean Ethics*, producing writings that supported Aristotle's conception that worth had its foundation in human demand/want/need. But they went into realms not addressed by Aristotle, arguing that the worth of an item was also dependent upon the labor and expenses used to produce it. Understanding how these two different views were reconciled by Albert and Aquinas remains of great historical interest. Though crude in detail, they had introduced the two primary elements of modern economic theory—demand *and* supply—into a single, unified discussion of the just price.

More important for the discussion of modern pricing, however, were two points the theologians and legal scholars ultimately agreed upon. First was the notion that *profits are just to the extent that they are based on costs, including the cost of labor and risk, and noble intentions.* Only excessive profits were sinful. From this observation, it might be concluded that a method for calculating the just sales price was at hand: determine the costs and account for noble intentions. The problem with this approach was that it could arrive at a price that was vastly different than the price observed in the market, posing all sorts of practical problems. Among other things, courts would find themselves involved in the practice of trade more frequently than they could support or that they probably felt was warranted. Legal scholars understood from practice what Albert and Aquinas were beginning to grasp from a philosophical perspective. The market price was determined by more than the cost of supplying a good. It was determined by the forces of demand and supply.

The second important point agreed on by the Church theologians and legal scholars was the method for estimating the just price in practice. Historical records show differences in the details, but the underlying theme was consistent. To Albert the Great, "a price is just which can equal the goods sold according to the estimation of the market place at that time." Simon of Bisignano (11??–?) referred to the "true value of goods as the price for which they were commonly sold." And Bernard Botone of Parma (?–1263?) came to the striking interpretation that "a thing is worth as much as it can be sold for." Perhaps my father's late night response to his inquisitive son's query about what something is worth was right after all.

It is important to clarify what was meant by market price. It was recognized that prices could and would vary from time to time and place to place. If grain cost a certain amount now in a certain city, there was no requirement that the just price be the same six months in the future in a different season, nor was there a requirement that the just price be the same in a different city. It was also recognized that there might be some variability in sales prices for the same or similar goods. For this reason, the just price, when needed, was calculated using variations on the same basic approach we use today: comparable sales were reviewed by a disinterested party or parties of good moral character, and a price was assigned that reflected these prices. Justice hinged on the presumption that *at a given place at a given time, a just price exists, and it can be reasonably approximated from price information in the market.*

* * *

The year 2005 would prove to be the worst hurricane season in recorded history. With 26 named storms, breaking the previous record of 21 dating back to 1933, the National Weather Service was forced to designate five storms with letters from the Greek alphabet: Alpha, Beta, Gamma, Delta, and Epsilon. Thirteen of the storms reached hurricane status, surpassing by one the record set in 1969. Three of the hurricanes reached the highest possible ranking of category five, indicating sustained winds of 155 miles per hour or more. Two of these category five storms—Katrina and Rita—made their way through the Gulf of Mexico before passing over the refinery-laden coasts of Texas, Louisiana, Mississippi, and Alabama. Damage from Katrina, which will long be remembered as the hurricane that devastated New Orleans, reached well into the hundreds of billions of dollars, making it far and away the most expensive hurricane in U.S. history.

As a result of these storms, oil and natural gas production in the region plummeted. From August 28 to September 2, the effects of Rita were felt as over 75 percent of the oil productive capacity along the Gulf Coast was out of service. Natural gas production fared even worse, with a figure of 85 percent. Following a brief respite in mid-September, the arrival of hurricane Rita on September 24 once again sent production into turmoil. For over a week, oil productive capacity was down by 75 percent. Natural gas production shut down entirely. The return to normalcy was slow. Not only were production facilities damaged, but so was the infrastructure designed to deal with such disasters. With 29 percent of U.S. oil production and 21 percent of natural gas production coming from the Gulf of Mexico, supply within the United States fell.

Not surprisingly, retail gas prices rose, and rose rapidly. From an average of $1.79 per gallon at the beginning of hurricane season on June 1, prices reached $3.08 on September 5 following Katrina. Prices would drop to $2.78 in mid-September, rising again to reach $2.97 on October 3 in the wake of Rita. Prices then began a gradual decline, reaching $2.17 at the end of November.

Energy company profits went through the roof as reported in the *New York Times* on October 28. Third-quarter profits showed extraordinary increases over the previous year: Exxon Mobil was up 75 percent on profits of $9.92 billion, Royal Dutch Shell was up 68 percent on profits of $9.03 billion, and Marathon Oil was up 347 percent on net income of $770 million.

Public outrage toward oil companies was severe. In a CNN/USA Today/ Gallup poll taken in late October, in which respondents were asked, "Do you think Congress should hold an investigation into the profits that oil companies have made in the past few months, or do you think an investigation is not necessary?" fully 82 percent responded that Congress should investigate. An ABC News poll airing on October 28 reported that 72 percent of Americans believed oil companies were taking advantage of the horrible season of hurricanes to drive up gas prices. According to one interviewee, "I don't think it's wrong that they're making a profit because that's the American way. But these sound like very extensive profits, and that's what doesn't seem fair."

How did the oil companies wind up in such disfavor with the American public? The rise in prices isn't difficult to explain with the most basic principles of a free market economy, a system that we go through great pains to defend and support. It would seem that if we accept this system, we should find no injustice in the prices it determines. The answer lies in our fundamental notions of justice.

First, while the market price was rising in response to the influences of both demand and supply, our sense of justice wanted to see oil companies receive a reasonable profit, something attached to the costs of production. With oil company profits soaring, it was difficult to see how the profits could possibly be reasonable. Just as in the Middle Ages, we felt injustice based on excessive profits and a sense that the merchants were having their way with us.

Second, prices changed so quickly that at a given place at a given time there existed no just price. What were we to compare the price at the pump with? Prior to the $1.29 rise that occurred in the three months from June 1 to September 5, 2005, prices had shown slow but regular growth of $0.34 for the prior 17 months. Had we continued to experience a slow rise in price, or a small jump that then remained relatively stable for a period of time, we could have established a referential or just price for comparison. With a rapidly rising price, our constant reference point was simply "something lower than it is now."

To understand our discomfort with airline ticket pricing, consider the notion of a just profit. The cost of running an airline is, to a large extent, fixed in the short to medium term. There are traditional fixed costs—capital equipment and labor—but even traditional variable costs such as fuel are often treated as if they are fixed. The reason is that airlines, like many businesses, want to keep their expensive capital assets generating revenue. Rarely do they undertake marginal profit calculations, looking at whether marginal revenues exceed marginal variable costs, to determine when it makes sense for an aircraft to sit idle. Instead, planes are expected to fly a certain number of hours each day, and marketing, route-planning, scheduling, pricing, and revenue-management departments are expected to generate whatever revenues they can.

As a result of how costs are conceptualized, it becomes difficult to intelligently ascertain what the "fair and reasonable price" for any flight itinerary should be. If costs are fixed for all practical purposes, then there's no compelling reason to charge based on distance flown other than that it appeals to our sense of justice because we connect flight length with cost. Why should a nonstop flight from Houston to Los Angeles cost $178 when a nonstop flight to Fayetteville, Arkansas costs $811?

Even assuming cost could be easily allocated among different flight itineraries, competition still factors heavily into the equation. If a competing carrier enters a market with much lower fares than those of an established network carrier, the network carrier might respond by setting a competitive price. In turn, the loss of revenues would need to be made up elsewhere in the network, and prices would become quickly misaligned with whatever measure of cost was adopted.

What airlines have done is to completely set aside cost when establishing the price that's available in the market. Mathematical models look ahead at how many passengers are expected to arrive and what they're willing to pay, and then set the available price based on an evaluation of each seat's potential revenue contribution. In a very real sense, airlines are adhering to Aristotle's conception that price is determined solely by demand, leaving cost out of the pricing equation altogether.

Pricing based on market willingness-to-pay is not a new concept and is generally considered a much better approach than cost-based pricing. It's also what differentiates science-based pricing from accounting-based pricing. Still, it offends our sense of justice even as we accept the workings of a free-market economy.

Beyond the effect on our innate need to see profits somehow connected to cost, there's an even more basic threat to our sense of the just price posed by airline ticket pricing practices. Scholars of the Middle Ages came to rely on the fact that *at a given place at a given time, a just price exists, and it can be reasonably approximated from price information in the market.* But if airlines are presenting us with constantly changing prices, what is our reference point for calculating a just price? Is it today's $300 or tomorrow's $400? Should we feel happy that we're saving $100, or angry that if we don't act now, we'll have to pay an additional $100? Adding to our sense of discomfort is uncertainty about what tomorrow will bring. To feel a sense of justice or injustice, we need a price that allows us to orient ourselves. By constantly changing prices, airlines have taken that away.

Airline ticket pricing is similar to an auction. As passengers, we aren't placing bids for specific flight itineraries (although auctions of various types are used, they account for a small number of ticket sales), but rather, from our purchasing behavior, we are making our preferences known. Airlines, in turn, use this information to make their own bid, setting a market price and asking if we're willing to outbid it. In fact, the range of emotions we experience when purchasing an airline ticket is often closer to that of a bidder in an auction than to that of a customer in a retail store.

* * *

As science-based pricing evolves, it must continue to keep the public perception of justice paramount. One of the great gaffes in public-relations history occurred in 1999 as the Coca-Cola Company experimented with vending machines that altered price based on temperature. Holding true to free market principles,

Chairman M. Douglas Ivester made comments in a Brazilian newsmagazine that would return to haunt him on the front page of the *Wall Street Journal* on December 17, 1999, in an article entitled "Tone deaf: Ivester has all the skills of a CEO but one: Ear for Political Nuance." Said Ivester, "Coca-Cola is a product whose utility varies from moment to moment. In a final summer championship, when people meet in the stadium to have fun, the utility of a cold Coca-Cola is very high. So it is fair that it should be more expensive. The machine will simply make the process automatic."

For a consumer-product company like Coke, image is unimaginably important in building and maintaining sales. One of the most valuable brands on the market today, Coke has an advertising budget that annually runs into the billions of dollars. Even if the company was justified in varying price based on the utility to customers, there was no way to avoid the public perception of price gouging.

Of course, there were many ways Coke could have avoided the situation. Perhaps the most effective were those that company spokespeople sought to emphasize in a *New York Times* article entitled "Coke tests vending unit that can hike prices in hot weather" that appeared on October 28, 1999. For one, rather than varying price based on temperature, the company could experiment with boosting sales during periods that are typically slow, for example, in office buildings during the night or on weekends. Alternatively, Coke could use detailed records of what sold and what didn't sell at a particular vending machine to determine how to better stock it in the future. It isn't difficult to imagine increasing revenues at a particular machine by a few percentage points just by getting the proper mix of product. Pricing would enter the equation if bottles of different sizes sold for different amounts, or if different products—Coca-Cola and bottled water, for example—sold for different prices. Taking a lesson about inventory management from the airlines, a vending machine could simply stop selling low-priced items during times of high demand, even if there were still bottles physically available. There are many ways to vary the price of a good without infringing upon the public perception of justice.

Setting aside the perception of justice, are these many new methods of varying price just in actuality? From the perspective of business, yes, in a free-market economy these methods are unequivocally just. Businesses create products and are free to sell them wherever they choose at whatever price they choose. There are, of course, exceptions. Laws exist to prohibit profiteering during periods of emergency or disaster. Anything that impinges upon competition—from collusion to monopoly power—is dealt with severely in U.S. courts of law. But these are not the situations in which airlines and other businesses are employing science-based pricing. They are seeking to do so in their standard, day-to-day sales activities.

At the root of the new pricing activities is an effort to find out what the market is willing to bear at a given time and place, and to charge a price consistent with market conditions. In concept, nothing has changed other than the level of detail and sophistication at which this goal is pursued. The consumer safety nets are those that have always been in place: competition and our right as consumers to say no.

There's a third net, however, that businesses need to remain keenly aware of: the consumer's sense of justice. It's not an entirely new problem, but science and technology are constantly developing novel ways for businesses to foray beyond the boundaries of generally accepted practices. Temperature-dependent pricing of bottled beverages was deemed unjust by the court of public opinion. Receiving discounts at the grocery store in exchange for information about our detailed purchasing habits has gained wide acceptance through the use of discount cards ("Thank you for shopping with us, Dr. Boyd"). We no longer require individual items to carry a price tag on store shelves, instead allowing tags to be placed on shelves near the items. Will we one day accept electronic price displays that can be changed with a signal from a computer? Will we ever allow differential pricing based on who we are? It's impossible to know exactly what activities will be viewed as just, and companies will need to chart the waters carefully as they enter new realms. But it's absolutely certain that, just as businesses will discover new ways to price, we as a society will make our position known about what we are and aren't willing to accept.

* * *

Though the moral obligation to set the just price no longer lies at the heart of our economic system, in their quest to come to terms with the just price, scholars from antiquity through the Middle Ages helped us better understand our deeply held beliefs about pricing. Airlines pushed the boundaries of generally accepted pricing practices and, in doing so, challenged our sense of justice in at least two ways: they developed a price structure that seems disconnected from actual costs, and they changed price with such frequency that we lost a point of reference allowing us to determine if, in fact, we were paying a just price. Our discomfort has mitigated itself as we've grown more experienced with the way airlines price, but it still remains.

I often pose the following thought game to people who advocate a single, unchanging price for airline tickets. Suppose you operate a one-plane airline that flies between New York and Miami once per day. One-way tickets sell for a fixed price of $300. However, on one particular flight departing 30 days in the future, natural statistical fluctuations in the number of booked passengers have left only one seat available on the trip departing from New York. You are all but certain there is someone out there willing to pay $500 for a ticket. As an airline seeking to remain profitable and stay in business, do you raise the price or not?

Many people say no, arguing that the damage to the airline's reputation in the long run is certain to exceed the additional $200 the airline will pocket now. If raising the price is a one time event, this assessment may well be true. But what if people learned over time to expect that such price changes might occur? Our $500 passenger might not be happy, but will he alter his buying habits? Perhaps the higher price helped him get the seat he wanted by protecting it from individuals with a utility measuring less than $500. Would the inability to get a seat if the airline didn't raise its price cause him to look to other carriers first? What if, as is often the case in real life, profit margins are frighteningly small? Most people quickly see the complexity of the question. But there are those who steadfastly claim that they would not change the price.

What would you do?

Notes

- Augustine's statement that "just as art cannot exist without imposture, neither can business exist without fraud" is taken from Baldwin's reading (1959, p. 15) of Augustine (*Ennaratio in Psalmum*, XXXIII, *Patrologia Latina*).
- The quotation from Tertullian (*De idolatria*, XI, *Patrologia Latina*) can also be found in Baldwin (1959, p. 14).
- Augustine's discussion with the merchant is taken from a summary of *Ennaratio in Psalmum*, LXX, *Patrologia Latina* by Baldwin (1959, p. 15).
- Vikas Bajaj. Oil companies report surging quarterly profits. *New York Times*, October 28, 2005.
- Detailed references for the practical interpretations of the just price arrived at by Albert the Great, Simon of Bisignano, and Bernard Botone of Parma can be found in Baldwin (1959, pp. 54, 71).
- Statistics on oil and natural-gas productive capacity were calculated from a report by Gilmer (2005) at the Dallas Federal Reserve.
- Gasoline prices for all grades of conventional retail gasoline were taken from government statistics found at http://www.economagic.com/emcgi/data.exe/doewkly/day-mg_tco_us.
- A transcript of the ABC News report, entitled "Mellody Hobson: Hope for Gas Consumers," can be found at http://abcnews.go.com/GMA/MellodyHobson/=story?id=1258388.

CHAPTER 9

The Scientists

Rumor is that Las Vegas won't have them anymore. As a group, they don't gamble enough. When they do gamble, they don't lose enough money. When they play blackjack, one of the few games in which a player can actually turn the odds against the house, pit bosses watch warily for any of the telltale signs that they're mercilessly taking advantage of an irregular run of cards.

The group is INFORMS, the Institute for Operations Research and the Management Sciences. Not every city is capable of handling their annual meeting, which hosts an average of 3,000 attendees. The problem isn't just the number of people but also the fact that most of them will take the podium at one time or another. Over a period of four days during the 2005 meeting, 16 time slots were set aside during which 50 simultaneous sessions took place, each session hosting an average of three to four speakers.

What do these people talk about when they get together? A sample of the presentations gives some hint: "Mathematical modeling of dynamic breast screening policies," "A staffing decision methodology for TSA security checkpoints at airports," "Capacitated production planning with price-sensitive demand and general concave revenue functions," and "Robust Wardrop equilibria." Topics range from the exceedingly practical to the intensely theoretical.

With such a wide variety of interests, subdivisions have formed. In all, there are 37, including subdivisions devoted to aviation, e-business, health, railroads, telecommunications, manufacturing, probability, optimization, the military, and even sports. One of the fastest growing subdivisions is the revenue management and pricing section. Spawned by the early success of the airline industry, this section serves as home not only for airline revenue management but also for the many pricing applications sprouting up in new industries.

What binds this eclectic group is a common tie to *operations research*, or simply O.R. for short. Arguably the worst name ever adopted by a group of professionals, it's not exactly about operations or research. The fact is, the discipline has never been able to escape its history and come up with a name that evokes a sense of what it's about. "Philosopher," "electrical engineer," "economist," and most other professional titles conjure up an immediate sense of what a person does. "Operations researcher" doesn't even slip easily off the tongue. The name *management science* is sometimes used, but is considered too restrictive by many since the discipline's focus isn't limited to managerial problems. And as a marketing firm charged with raising the visibility of INFORMS recently concluded, *operations research* carries more brand equity than *management science*.

The name is only the beginning of the discipline's difficulties. Describing what operations researchers do is equally difficult. On one hand, operations researchers are fundamentally problem solvers. They like to take complex problems, reduce them to their basic components, and come up with optimal solutions. Unfortunately, while this description is fundamentally correct, it's far too vague. Anyone who works or runs a household for a living would likely describe himself as a problem solver, yet very few actually practice operations research.

Sometimes operations researchers avoid describing what they do and simply point to the results of their work. When checking in at the airport, renting a car, or mailing a package at the post office, we usually stand in a single long line and go to the first available server when we reach the front. This wasn't always the case. Lines used to form behind each individual server as they still do at grocery stores. The single line reduces the average waiting time per customer substantially, and was due to the work of operations researchers. The problem with such descriptions is that they don't convey any sense of what lies in the background. A listener may well walk away with a sense that operations researchers are efficiency experts and, as I was once informed, do work that is "dangerously close to common sense."

The reality is that operations research is a form of mathematics. While problem solving is a vast topic, it turns out that the mathematical tools employed to solve most quantitative business problems aren't as numerous as might be imagined. Random events are an important part of many problems, whether they correspond to the length of time people spend waiting in line, the number of casualties sustained in a military action, or the number of people buying plane tickets. In the parlance of mathematics, these types of random events fall under the heading of *stochastic processes*, and comprise one of the two major branches of operations research.

The other major branch is *optimization*, which deals with finding the best solution to well-defined problems. A good example is that of finding the shortest route between any two points in a city. Given a list of the streets and their length, there are optimization methods that can find the shortest route between any two points in a mere fraction of a second. If a computer actually had to try all the different routes and compare them, then for any city of even moderate size, today's computers would take longer solving the problem than the estimated age of the universe.

Operations researchers receive formal training in both optimization and stochastic processes, normally including some training in statistics, forecasting, and computer simulation of complex systems. They also receive training on how to use these tools to solve real problems.

The fact that a common set of mathematical tools can be used to address such a wide variety of practical problems is nothing short of remarkable. Abstraction is the key, recognizing that orders arriving at a warehouse and people calling to purchase airline tickets can be described in fundamentally the same way. While the actual business problems are quite different, when abstracted, they can be studied together.

Yet the efficiency provided by abstraction is one of the greatest challenges facing the discipline. A truck dispatcher for a delivery company isn't likely to see the connection of his work with designing printed circuit boards or doctors using radiation to destroy tumors. There should be people who study truck dispatching, and they should belong to a truck dispatching professional society, just as printed circuit board designers and radiation treatment professionals should belong to their own societies. To everyone who isn't an operations researcher, problems are defined by problems, not a common, abstract mathematics. This is why most operations researchers develop a close affiliation with a profession other than operations research. More than a few leave the discipline behind. Still, many cling to the mathematical foundation that helped draw them into operations research to begin with, and this is why they return year after year to a common meeting place.

The allure of mathematics, however, is a two-edged sword. While mathematics attracts people to the discipline of operations research, it also creates a barrier between those who develop mathematical tools for activities like pricing and those who actually use the tools. Operations research is tailor-made for today's world, a world filled with data and technical challenges. But getting industry to accept operations research requires communication in nonmathematical terms: it requires communicating what the discipline is and, at a deeper and more important level, communicating the discipline's value to a world that isn't mathematical. The extent to which scientific pricing will be embraced by industry depends on operations researchers' ability to tackle the many nonmathematical pricing challenges they face with the same passion they devote to the mathematics of pricing.

* * *

When I sat down with Thomas Cook, I'd known him for many years, both as a competitor and a respected professional colleague. But sitting with him for a day to talk about his past life and future plans provided an opportunity to better understand him. A recent president of INFORMS, Cook knows the importance of communication as well as anyone. Holding an undergraduate major in mathematics from Grinnell College, an MBA from Southern Methodist University, and a PhD in operations research from the University of Texas, Cook spent six years teaching, first at the University of Tulsa and later at Boston University. But he soon realized that university life wasn't his calling. After three years consulting with Arthur Andersen, Cook found himself director of operations research at AMR, the holding company for American Airlines.

When Cook came on board, his group consisted of eight people. Jeff Katz, future CEO of Swissair and Orbitz, was among those who worked for Cook. As Katz rose through American, he would smilingly refer to Cook and his group as "those hoodlum PhDs."

The group Katz was referring to was one of the fastest growing in the company during a time when most other groups in the organization were losing staff. Cook was a good salesman. Coming on board, he was quick to transform the operations-research team from its research focus to a concentration on doing projects of direct, measurable value to American. Cook knew how to demonstrate his group's value and how to take advantage of growth opportunities when they presented themselves. But he also had a sympathetic ear in CEO Bob Crandall. Crandall, who always viewed himself as a leader in the use of technology, took little convincing that operations research was a good investment. After Cook experienced some early successes with his operations-research group, it was, in fact, Crandall who initiated the group's expansion. Calling Cook into his office, Crandall informed Cook that operations research was so important that he wanted to see a growth plan. Cook prepared a proposal as requested and presented it in front of Crandall and the senior management team. Cook recalled, "With everybody cutting back staff, Bob was probably the only one in the room that liked it. But when I was done, he looked at me and said to go for it before turning to everyone else at the table and asking, 'Anybody have any problems?'" With that, Cook had his mandate to build a bigger, more prominent operations-research group.

But before he accepted, Cook went back to Crandall with one request: he needed freedom to hire foreign citizens without concern for their visa status. The U.S. government not only sets limits on how many people are allowed into the United States from foreign countries, but companies that hire too many foreign citizens are also subject to investigation. In the mid-1980s, Cook's issue wasn't that he wanted to outsource. The problem was that he knew he couldn't find enough qualified U.S. citizens to fill the vacancies he was faced with.

The problem wasn't limited to operations research and it wasn't limited to the 1980s. Following a recruiting trip to Texas A&M, one of the two major public universities in Texas, I was presented with 21 resumes, nearly all of which were worthy of an interview. Unfortunately, 18 were from Indian citizens and three were from Mexicans. Legally, my hands were tied.

The situation begs the question why universities in the United States don't admit more U.S. citizens. In reality, science and engineering schools clamor for good U.S. citizens. Whereas a student entering the law program at the University of Texas might expect to pay $15 thousand per year in tuition, a student entering the graduate engineering program might expect to have tuition waived and receive payment of $15 thousand per year or more.

The problem is not so much that foreign citizens with slightly better scores on standardized tests are granted admission, but rather that there aren't enough U.S. applicants with scores high enough to indicate they can successfully complete a graduate degree. Often, there simply may not be enough U.S. applicants to fill out enrollment even if qualifications are completely ignored.

Statistics for doctoral programs make the situation frightfully clear. For the 2003–2004 academic year, the Duke University Department of Electrical and Computer Engineering received a total of 323 applicants, of which 20 were U.S. citizens and 303 were foreign. The department made 31 offers, with 15 going to U.S. citizens. Ultimately, 19 students entered the program, 7 of whom were U.S. citizens. The training of foreign citizens isn't an issue, since they typically perform as well as or better than their U.S. counterparts. However, due to cultural and language problems, their ability to communicate in business situations is frequently hampered. In spite of potential communication problems, Cook knew it would be impossible to grow his group without the help of foreign talent.

With Crandall's okay to hire foreign students trained in the U.S., Cook was on his way to building the large, important organization he envisioned. Five years after starting with a group of 12, Cook was leading some 75 people looking at all aspects of airline operations. Revenue management remained one of the most important areas of activity, since Crandall not only understood its value, but also was ultimately responsible for making it happen at American. As revenue management at American evolved, Crandall certainly wouldn't keep up with the details in the science, but he understood the fundamentals, and more than anyone, he understood the power of revenue management as a competitive weapon. "Bob was always the smartest guy in the room," said Cook.

Cook was as successful as anyone in communicating with Crandall, but he wasn't immune from mistakes as he learned during one particularly rocky encounter. Delays were a regular topic at Crandall's weekly staff meetings, a topic known to produce casualties as Crandall made his intentions all too clear for those around him. Delays not only affect passengers and the smooth running of an airline's interconnected network of flights, but they also affect statistics. Airlines measure their performance by statistics and use statistics to compare their performance against other airlines.

In running an airline, as in life, there are events that can and can't be controlled. Maintenance is a controllable event. If a plane is delayed because it wasn't serviced properly, an airline can take steps to rectify the problem in the future. Delays experienced as a result of ground holding imposed by air traffic control are not. Cook realized that all the factors impacting delays were being captured in the same statistics, and during a staff meeting, proposed that delays be broken into controllable and uncontrollable events and the statistics reviewed separately.

Rather than seizing the moment with an insight Crandall would embrace, Cook quickly learned that he had blundered. Delays made Bob Crandall unhappy and the source was irrelevant. When Cook confronted him with the most obvious example of an uncontrollable event, the weather, Crandall's response left no doubt as to where he stood. Slamming his fist to the table and yelling at the top of his lungs, Crandall reminded those present that for American Airlines, "There *is* no such thing as an uncontrollable event."

Crandall isn't unique. Business leaders—people who have the strength to run a large, profitable organization—learn to count on themselves. Right or wrong, when they make a decision, they need to believe they're right and communicate this to their organizations. Part of what comes with this mindset is that they'll listen to those around them, but only when they're damn good and ready. Those so fortunate may receive a brief, scheduled meeting, but as leaders grow busier and operate further up the corporate ladder, even this level of communication becomes less likely. Lunch, a chance meeting in the hallway, or staff meetings may be the only opportunity to get a point across.

Venture capitalists have a term for the type of communication required in these situations: the *elevator pitch*. Stepping into an elevator, an entrepreneur finds himself face-to-face with a representative of the venture capital firm of his dreams. He has from the time the door closes to the time it opens to make a convincing argument for his company to receive $10 million in funding.

Communication on the order of an elevator pitch is more frequent than we allow ourselves to believe. Good resumes are no more than two pages in length, yet seek to distill a person's entire professional career. Book authors, capable of writing hundreds of thousands of words on a topic, have a page to capture an editor's attention with their query letter. Yet every editor knows that even the query letter is too long. The "high concept" needs to sell the book in the first few sentences.

Certainly, not all communication takes place in such an abbreviated fashion. Further down the corporate ladder, managers have more tightly focused responsibilities and engage in longer, more detailed discussions. But the ability to communicate quickly and effectively is always, always important. No matter who is listening, their attention span is limited, and the less interested they are in a topic, the shorter that attention span is. They also come to any discussion with their own knowledge and background, not to mention their immediate concerns of the day, whether they had a bad morning send off from their spouse or a developing soreness in the back of their throat.

People with technical backgrounds have a hard time appreciating the need for brief and effective communication. For one part, they simply aren't trained to think this way. Complex problems are by definition complex, and need to be treated as such. Technical people also tend to assume that those they speak with know more than they actually do. Often, their convictions lead to frustration as they seek to resolve misunderstandings by talking more and in greater detail.

One of the activities at the INFORMS annual meeting is the Doctoral Colloquium, where top students from universities around the country are invited to spend a day learning about careers in operations research. Speakers include recent graduates along with well-established figures in the field.

During one meeting, a young faculty member was speaking about the process of giving a research presentation as part of an interview, a requirement for academic positions. After outlining what to do with the first 30 minutes of a 50 minute talk, the speaker described how to handle the final 20 minutes. In unambiguous terms, the speaker made it clear that no one should understand the remaining material. The last 20 minutes weren't for clarification, but rather to demonstrate intellectual superiority in a narrowly focused domain. There was no disagreement from anywhere in the room.

Taking the podium immediately following the speaker, I couldn't help but comment on what I'd just heard. Potential young faculty members were being told to communicate so that no one could understand what they were talking about. *And I couldn't disagree.*

Especially in fields with a significant mathematical component, it's not sufficient to simply demonstrate that research breaks new ground. The work also needs to be hard. As the speaker pointed out, if someone grasps the details of whatever is being presented, you risk being accused that the work is too easy. Presenting material that is over the heads of the audience is effective communication in this instance, since it achieves the speaker's goals and is well within the norms of what the audience expects.

The unfortunate consequence is that the academic world of operations research rewards actions that are completely at odds with what is required to be successful in business. For some disciplines, this is acceptable. Electrical engineers who want to remain practicing electrical engineers can carve out successful careers in the mold cast for them during their university training. But for operations research, a discipline based on using mathematics for solving real business problems, failure to develop proper communication skills is devastating.

Leading a group of scientists, half holding PhDs and the other half holding master's degrees, I'm constantly drawn into problems stemming from communication. If men are from Mars and women from Venus, scientists and businesspeople are from parallel universes. Facilitating communication is like looking for a wormhole between these strange landscapes, then prying it open just long enough for each person to catch a glimpse of the other person's world. The psychedelic imagery used by Stanley Kubrick in the film *2001: A Space Odyssey* often comes to mind, complete with the obvious fear on the face of astronaut Dave Bowman on his sojourn through the universe.

In one instance, PROS was in the process of replacing a first generation airline revenue management system with a far more advanced system. The carrier would experience higher revenues as a result of the fact that more advanced science was being used to calculate what fare classes were available at any given time. When the system was turned on, the client asked a reasonable question: how were the new number of seats assigned to each fare class going to generate more revenue than before?

In fact, answering the question was complicated. In purchasing the new system, the airline was no longer looking at individual flight legs one by one, but was rather taking into account every flight leg in the entire network. Seat availability on every leg depended upon all the complicated interactions of passengers flying every which way. This is why the carrier had purchased the system to begin with.

When day after day passed without the client getting a satisfactory answer, the issue found its way to Bert Winemiller, PROS's CEO, who in turn pulled me aside. Winemiller, who holds a master's degree in statistics in addition to a

Harvard MBA, had caught wind of the fact that the PROS team had answered with a combination of mathematics or, when that didn't work, falling back on the time-honored answer "because." He was emphatic that there had to be a way of explaining what the system was doing so that anyone could understand.

A seemingly impossible task to many scientists, the answer lies in what's meant by an explanation. To a scientist, an explanation consists of understanding the mathematical reasoning behind how seat availability is calculated. To a businessperson, an explanation means why, from a business perspective, the new solution is better than the old. To this end, a good example is a much better way to explain a concept than a set of equations. Even better is to explain that *this* flight leg has overall higher prices because of the demand from connecting passengers on *these other flight legs*; that the cheapest fare class from Atlanta to Miami is $200 more than the competition because the competitors don't have a lot of people wanting to fly from Seattle through Atlanta to Miami who are willing to pay $700. Many problems can be resolved by understanding what the other party is looking for in an explanation and providing an explanation that makes that person happy.

For a scientist fresh out of school, communication mistakes are understandable. Surrounded for years by people who live and breathe a special language, it's natural to develop a perspective that the rest of the world speaks that language, too. As a hands-on personality, Winemiller finds himself talking to people up and down the PROS organizational structure. On one occasion, a new scientist no more than a month out of school happened to be in an important project review led by Winemiller. Asked a very basic science question by the CEO of his company, the young man rose to the occasion. Looking Winemiller in the eye, he proceeded to give an answer that must have come straight from his doctoral dissertation. Winemiller questioned the young man further, but when it was clear that he wasn't going to get an answer that satisfied him, he didn't press further. Sitting with Winemiller immediately following the meeting, he turned to look at me. Head shaking, a smile broke out across his face. "Don't ya just love these kids? He's a keeper. I know you'll have him back to earth in no time."

In the right environment, most scientists can learn good communication skills. But there are always those who, for one reason or another, never seem to make the transition from the world of mathematics to the world of business. Some have difficulty seeing the world through nonscientific eyes; some simply don't want to. There are those who view good communication as a sign of weakness, diminishing their reputation in the eyes of their peers. Others live in fear that they will somehow be discovered as an inferior intellect if they don't hide behind mathematics. Still others view educating the heathens as a calling, and are willing to fight to the death to make themselves understood on their own terms. Worst of all are scientists who understand the importance of communication but dismiss it. One of the most offensive comments I've ever heard, trivializing the efforts of managers, pricers, sales agents, and software developers in a single breath, was expressed by an especially dogmatic scientist. "Mathematics is what's important," he said. "Everything else is just the plumbing."

Whatever the reason, poor communication has been an important factor in thwarting the rapid acceptance of science in the day-to-day operation of business. Industry has only begun to experience the power of operations research, and it has rarely experienced it near the upper reaches of management. The extent to which operations research will be able to embed itself in the mindset of organizations in the future is directly related to the ability of those who practice it to communicate effectively at the highest levels of business.

Cook is one of the few people who understands this. Having built his group to 75 people, he realized future growth opportunities within American were limited. So he went to Crandall with the idea of starting a new entity under the AMR banner that would sell and market American's problem solving expertise, primarily to other airlines but also to other industries. At first Crandall wasn't receptive. After all, why should American sell its own highly sophisticated technology to its competitors? Cook, for his part, was persistent. He was able to convince Crandall that American would always stay ahead of other carriers through its ongoing research activities. But there was also the long held industry dictum that "you're only as smart as your dumbest competitor." Crandall had been through the years following deregulation when upstart airlines flooded the market with low fares, demonstrating little understanding of the long-term costs of running an airline. Competitor that he was, Crandall was willing to fight like hell with anyone who took him on. Yet wouldn't it be easier to simply deal with a smart competitor?

Whatever Crandall's reasoning, Cook got his way and American Airlines Decision Technologies (AADT) was born. Cook recalls that the early days were tremendously fun. There was the excitement and enthusiasm of doing something new, of selling and winning, of meeting the technical and mathematical challenges presented by the many problems he and his group faced. Remembering the moment when he received word of AADT's first big contract, a contract to work with Amtrak, Cook recalls letting loose a Texas holler that could be heard throughout the building.

Keeping with his early philosophy that operations research was about doing projects of direct, measurable value, Cook built a highly entrepreneurial environment. Consulting and product teams became individual profit centers. Five years after he started AADT, Cook was able to look at a group of roughly 600 people, almost half of whom had degrees in operations research. According to Cook, it was the largest group of operations researchers ever assembled in a single organization other than the U.S. government.

In the mid-1990s, AMR began the process of separating its core holding, American Airlines, from its SABRE reservations system and related technical services. The new corporate entity going under the name of the SABRE Group would ultimately achieve complete independence from AMR and become listed as part of the S&P 500. Along the way, Cook convinced Crandall to merge AADT with parts of the SABRE Group to form SABRE Decision Technologies, an organization of over 3,000 with Cook as its head. With SABRE Decision Technologies, Cook now managed a much more diverse collection of individuals.

Many of the new people reporting to him were information technology professionals used to working for a single airline, an environment quite different from the high-energy life of a growing entrepreneurial venture. Displaced from the work he enjoyed most, Cook would leave his position a few years later for a life of semi-retirement.

To this day, Cook steadfastly clings to his vision of operations research. It must bring demonstrable value to any organization it services. By focusing on value, everything else falls into place. But even as Cook rose through the corporate ranks, he never lost sight of the underlying science. He still enjoys talking about crew scheduling algorithms or the latest scientific trends in dealing with irregular operations caused by weather. His communication skills allow him to move easily between academicians and upper-level management, and these skills help him articulate his vision of the future of operations research to others.

With the support of Bob Crandall, Cook's efforts raised the visibility of operations research and scientific pricing to remarkable levels. Cook wasn't the first operations researcher in the airline industry, nor were he and his team the only people working on scientific pricing. The record of conference presentations and journal publications shows that operations researchers were slowly gaining interest in scientific pricing as early as the 1970s, and the late 1980s began a flood of published research. How operations research got its foot in the door was less a single event than it was a slow emergence. But Cook showed that operations research was a force to be dealt with by placing it squarely in the sights of executives throughout the industry. After all, if Crandall was making such a big investment in these "hoodlums" and experiencing such tremendous financial success in doing so, it was something to keep an eye on. Yet even as operations research was realizing a new kind of success in the business world, it was still dealing with its own internal problems.

* * *

Professional societies and other organizations form when people band together with a common set of goals. Environmental groups work to protect the environment, and chambers of commerce work to promote businesses within a local community. Because these societies are comprised of individuals, it's not surprising that they periodically argue about how to achieve the goals that unite them. In fact, a certain measure of disagreement is healthy and helps keep a society on track. Sometimes, however, there are fundamental conflicts that stand in the way of a society achieving its objectives.

For operations research, the problem is one of identity. From the beginning, operations research struggled with the balance between theory and practice, between mathematics and solving real problems. In May of 1952, 75 individuals from industry, the military, and academia met to discuss the formation of a new professional organization: ORSA, the Operations Research Society of America. The meeting was led by Philip Morse, who had headed the U.S. Navy's Anti-Submarine Warfare Operations Research Group during World War II. Momentum for the field was already apparent, and by the end of the year, the young organization would publish the first issue of its flagship journal and organize a national meeting attended by 400 members and guests.

A Princeton-educated physicist, Morse and the early leaders of ORSA were primarily interested in mathematics and military applications, or so it seemed to much of the early membership. Shortly after its formation, small, informal meetings sprung up around the country to discuss ORSA's direction. One such meeting was held in the Manhattan apartment of a Columbia University professor. A major complaint of the dozen or so participants was ORSA's lack of focus on how the newly evolving science could be applied to problems in management. Operations research had tremendous potential that was being overlooked.

However, there was another issue, as well. In 1953, Andrew Vazsonyi was working at Hughes Aircraft. Assigned to analyze problems related to factory production control, he stumbled upon ORSA and was asked to give a plenary talk at a society meeting. Hundreds of people were in the room to hear what he had to say. During questions at the end of the presentation, Vazsonyi was asked if he was a full member of ORSA.

"I knew what the heckler was getting at," said Vazsonyi in a publication celebrating 50 years of operations research.

When ORSA was founded, the leadership drove an elitist wedge into the association by creating two classes of members. Full members were only those theorists and mathematicians certified by the core group. On the other hand, associate members were riff-raff from the real world—guinea pigs from business who could try out the full members' theories, . . .

I decided to address the issue head-on. When the silence in the room became unbearable, I answered with as much humility as possible: "I am only a lowly associate member and do not belong to the inner sanctum." I received a standing ovation.

Eighteen months after ORSA was founded, a group of economists, engineers, mathematicians, statisticians, philosophers, and attorneys met in a room on the Columbia University campus to discuss a new entity that would rally around the name *management science*. Among those in attendance were four future Nobel Prize winners and a future recipient of the National Medal of Science. Thus it was that in 1953, The Institute for Management Sciences, or TIMS, was born.

For decades, the two organizations would live independent lives. By the end of the 1950s, ORSA boasted some 3000 members, with TIMS only slightly behind with 2600. The numbers were slightly misleading because many people maintained membership in both organizations. Both were focused on problem solving, but each had a reputation for a slightly different focus. Reviewing early articles from the two organizations' flagship journals, appropriately named *Operations Research* and *Management Science*, it's not immediately apparent that the two were as different as their reputations would have it. While *Operations Research* gave more space to military articles and *Management Science* to business problems, each used roughly the same level of mathematics and most of the articles could easily have appeared in either journal. What is clear is that as the years passed, both journals published longer, more detailed, more mathematical papers. Titles from the 1950s such as "The reliability of airborne radar equipment" (seven pages, two references) and "On bus schedules" (eight pages, no references) gave way in the 1970s to "Interior path methods for heuristic integer programming procedures" (19 pages, 16 references) and "A closure approximation for the nonstationary M/M/s queue" (13 pages, 28 references).

The 1950s and 1960s are remembered as a time of unfriendly competition between the two societies. But by the 1970s, the beginnings of a truce were in evidence. ORSA and TIMS jointly published the aptly named journal *Interfaces*, which sought to elevate the importance of practice. The two societies also began holding joint national meetings. Although both organizations would have their own president, board of directors, and business offices, for most of those in attendance, the distinction was never visible. People went to "ORSA/TIMS" meetings, and as new generations affiliated themselves with operations research, there was often confusion that only one professional society existed when in fact there were two.

After years of proposals and counterproposals surrounding a merger, on January 1, 1995, both ORSA and TIMS ceased to exist and were replaced by INFORMS. An important step in unifying the operations-research community, it didn't eliminate the many differing opinions on the theory-versus-practice debate. Practice, however, would find itself gaining momentum as dictated by the world around it.

Operations research curricula never completely found a home in universities. Many schools incorporated operations research in their industrial engineering departments. By and large, it was a fruitful wedding, and many departments took on the name Industrial Engineering and Operations Research. Business schools also began departments in operations research alongside accounting, finance, and other fields, though in many cases, the departments bore names that included "decision" or "information" and no explicit reference to operations research. Operations research showed up in mechanical engineering departments, mathematics departments, or no department at all. MIT, where Philip Morse spent his professorial years, operates an interdepartmental program with its own administration and students, but no formal faculty appointments. Instead, faculty are drawn from across the university, including the School of Management and the Departments of Electrical Engineering and Computer Science, Civil and Environmental Engineering, Ocean Engineering, Mathematics, Aeronautics and Astronautics, Mechanical Engineering, Nuclear Engineering, and Urban Studies and Planning.

The lack of a common home not only adds to the confusion over identity, but it also makes the discipline susceptible to losing university support if the world takes a new direction. History departments are safe since it's impossible to imagine higher education without the study of history. The same can be said of English, biology, chemistry, and other fields that have established departments in the university setting.

A huge blow was dealt to operations research with the change in accreditation standards for business schools in the 1991 report of the Association to Advance Collegiate Business Schools. The late 1950s had seen the growth of quantitative methods in business schools as it became clear that knowledge of basic mathematical principles was needed for managers of the future. Prior to 1991, business schools were required to provide "the equivalent of at least one year of work . . . in quantitative methods" and other closely related disciplines. Protected by these standards, operations research faculties were free to develop curricula as they saw fit. The result was that courses in mathematical algorithms were being taught to MBAs.

The 1991 report changed the guidelines, requiring only that the curricula for MBAs "should include instruction in . . . financial reporting, analysis and markets; domestic and global economic environments of organizations; creation and distribution of goods and services; and human behavior in organizations." Gone was the requirement that business schools teach a year of quantitative methods. An era of far greater latitude for business schools was underway, an era that would see students emerge as customers and business schools as businesses. In this environment, operations research professors were forced to make their courses not just relevant, but attractive to potential students. *Operations management*, which uses operations research as a tool but focuses more broadly on all aspects of procurement, manufacturing, and distribution, gained strength at the expense of operations research.

In engineering schools, where faculty are expected to draw money from outside the university, government research funding was slowly disappearing. Military agencies had long supported operations research, as had the National Science Foundation. But with fewer exciting, fundamental breakthroughs than in earlier years, funding agencies were turning their attention elsewhere. Coupled with the general decline of government research funding overall, engineering faculty found themselves increasingly seeking money from private enterprises, a source known for its interest in practice, not theory.

Within INFORMS, the Edelman Prize emerged as the most sought-after award by the membership. Referred to as the "Nobel Prize" or "Super Bowl" of operations research, the award recognizes achievements in the practice of operations research. To win the award, someone must actually use operations research to solve a problem, and the results must be clear. While the prize committee leaves open the door for inventive solutions that don't directly involve money (curing cancer would be given serious consideration), it's generally understood that to win the prize, it's necessary to save an organization vast sums of money and have senior management show up to attest to this fact. When top executives can't attend the competition, videotaped, professionally produced statements are accepted, but with more and more executives making time to attend the competition, videotapes carry less weight than in the past. General Motors was the award recipient in 2005, making the case that operations research saved $2 billion through improved productivity at ten manufacturing plants located throughout the world. The winning numbers aren't small.

The Edelman competition is the highlight of the annual INFORMS Conference on OR/MS practice. Designed in an effort to attract practitioners interested in learning more about operations research, the practice conference has supplanted one of the biannual meetings of the general membership. A creative idea, the meeting has yet to attract a wide audience beyond members of the operations research community.

Recognizing the demands being placed on the profession, INFORMS and its preceding organizations made great strides in turning the wheel back in the direction of practice. Still, there is no question that theory stands firmly at the helm, and for understandable reasons.

* * *

Garrett van Ryzin began his career as a rock guitarist. Fresh out of high school, he spent a year working clubs in his home town of Madison, Wisconsin, before moving to Nashville. "Living in a music town is sobering," said van Ryzin. "There's so much talent, and so many people that achieve medium success before completely disappearing." When the group he formed with his brother lost its fourth drummer, he decided it was time to go to college—Columbia for his undergraduate education, then on to MIT. He still keeps a small collection of electric guitars and prowls the Web in search of more when he's not busy writing scholarly books and papers.

Prematurely gray, van Ryzin parts his hair in the middle and keeps a tidy beard and mustache. He's usually found in a sports jacket and cotton pants. In a room filled with people, van Ryzin would easily be picked out as the academic. His mannerisms only serve to further this perception. He often rests his hand on his chin when thinking, and speaks in rapid sentences interspersed with short, clipped phrases as his mind sorts out where it's going next. His laugh is loud and frequent and cheerful, but short.

Now the Paul M. Montrone Professor of Private Enterprise at the Columbia University Business School, van Ryzin has spent most of his academic career concerning himself with pricing problems. Commemorating the fiftieth anniversary of *Management Science*, the editorial board chose what it considered the 50 most influential papers to have appeared in the journal. Among the papers was the 1994 publication "Optimal dynamic pricing of inventories with stochastic demand over finite horizons," coauthored by van Ryzin and his colleague Guillermo Gallego. Based on work with a garment retailer, it was one of the earliest works to take a quantitative look at the seasonal pricing of clothes, including markdown pricing and sales. The somewhat broader theme of retail pricing has since become one of the more successful applications of scientific pricing outside the travel industry.

One contribution of the paper is an emphasis on the formal link between dynamic pricing and what the airlines have done with revenue management. Van Ryzin would go on to hook up with former MIT classmate Kalyan Talluri, then at U.S. Airways, to write a series of papers on airline-style revenue management. Their work together would culminate in a 700 page book entitled *The Theory and Practice of Revenue Management*. In 2005, the book would win one of the oldest and most distinguished awards offered by INFORMS, the Lanchester Prize. At one time the most sought-after prize in operations research, it has been eclipsed by the Edelman Prize. Asked why the title of the book doesn't include the word "pricing," van Ryzin responds that he and Talluri consider pricing to be just one branch of revenue management as it is now coming to be defined.

As an Ivy League professor and editor of the journal *Manufacturing and Service Operations Management*, van Ryzin understands the requirements of publishing in top, peer-reviewed, academic journals. Academic circles place a huge premium on mathematical sophistication. No matter what is said about the importance of practice, junior faculty seeking to establish their reputation and earn tenure know that if they want to get published, they need to do good theory. In an effort to generate the required number of publications before the tenure clock runs out, junior faculty are often forced to work on problems that have only a loose connection with anything real. If the problem sounds real but yields interesting results that are hard to prove, it is much more likely to get published than a real problem that is solved by ingenious means but doesn't give rise to difficult mathematics.

To help resolve this discrepancy between publication and practice, some schools are providing young faculty with a leave of absence to spend time in industry before starting their university careers. Funding organizations such as the National Science Foundation are providing monetary support for professors to get out of their offices and into the workplace. Universities are now hiring young people who started their careers in industry but are looking to return to the university. Ten years ago, even in an applied field such as operations research, time in industry all but assured the academic career path was closed. Today, universities not only value the practical experience, but also the network of funding sources for the young academician's research.

The perfect paper solves an important practical problem and does so with difficult yet insightful mathematics, mathematics that demonstrates some fundamental new property about problem solving in general. Not surprisingly, such papers are few and far between. Recognizing this, van Ryzin's editorial policy is to accept papers that make a contribution in one area or the other. In addition, during his time as editor, he has instituted a new category for practice papers. Van Ryzin isn't the first to undertake such efforts. For a period of almost ten years starting in the mid-1960s, the journal *Management Science* published two separate series simultaneously, one devoted to theory and one devoted to practice. Other journals, such as *Operations Research*, also have a special section for practice papers. At present, the trend once again seems to be toward a growing acceptance of practice papers.

Theory isn't on the ropes, however. For many academicians working in technical fields, theoretical work is simply the most intellectually rewarding. There is nothing quite like the feeling of discovering a deep new understanding about the way the world works, even if the "world" is an idealized mathematical abstraction complete with frictionless wheels and spherical horses. This feeling, coupled with a desire to share their work with peers, keeps the best academicians up late at night. Even if only a handful of people will understand what they have to say, and even fewer will take the time to listen, creativity drives them to spend endless hours reading, thinking, and writing.

The vast majority of academic research will never find its way into practice simply because it has no practical relevance. Some ideas with practical merit will be lost because they are buried in technical jargon and never communicated outside a small group of like-minded individuals. While publication in the popular press is increasingly acceptable, academicians who overstep the line are subject to the scorn of their peers.

A small number of ideas will escape the academic circle and have a profound influence on the way we do business. The mathematical details of airline ticket pricing are but one example. Economist and operations researcher Kenneth Arrow received the Nobel Prize for extremely deep yet practical insights into the limitations of voting procedures. Proposing a very basic set of rules that every voting scheme should adhere to—for example, when everybody prefers Jim to Steve with the exception of one person, then Jim should be elected—Arrow then proved the stunning conclusion that the rules were mutually contradictory; that there don't exist any voting procedures whatsoever that satisfy all of them. Harry Markowitz received the Nobel Prize for formalizing the importance of risk when constructing a portfolio of investments. Extensions of his work now dominate how fund managers build and maintain their portfolios. Advances such as these arise from an environment that encourages theoretical investigation. In many instances, it's difficult to point to a particular instant when an idea took root, but it's not difficult to look back and see that somewhere along the way, the world changed. Without an adequate theoretical foundation, operations research would dwindle as it failed to address any big questions. It would cease to be a science.

* * *

For all the theory-versus-practice debate, there's no question among those involved with operations research that science equals money—so much money that operations research isn't just useful, it's critical. In 1998, Cook declared that revenue management generated nearly $1 billion annually in incremental revenues for American Airlines. Inevitably, as the competitive forces of a free market economy work their magic, companies without similar leverage will ultimately be driven out of business.

The question is not *if* science will continue to take root in business over the coming decades, but the *extent* to which it will take root. Pricing science faces special challenges since most pricing organizations aren't very scientific to begin with. This makes it crucial that those who understand the science take time to communicate with those who don't, recognizing that that listening, empathizing, educating, and even compromising are as important as mathematics. Though pricing scientists don't always realize it, the future of scientific pricing is dependent upon their ability to transcend the barrier between mathematical models and the real world's many nuances.

Notes

- Information on the history of SABRE is from an interview with Thomas Cook and from Cook (1998).
- Much of the history of INFORMS, ORSA, and TIMS is from Horner (2002). The quotation from Andrew Vazsonyi is taken from Vazsonyi (2002).
- The discussion of business school accreditation standards is from Grossman (2003a, 2003b).

CHAPTER 10

The State of Pricing

Working as a professor in an engineering school isn't the ivory tower experience most people envision. True, teaching can be very rewarding and nobody tells you what you're supposed to research. But unlike professors in law schools or schools of the humanities, engineering professors are required to attract money: money to pay for equipment, money to pay for student tuition and stipends, and money to pay for a reduced teaching load. All in all, it's a healthy arrangement, providing professors with motivation to work on problems of relevance, since there's a strong correlation between what people want researched and what they're willing to pay for. However, sometimes the job seems more like a position in sales than in teaching and research.

In one particular effort to attract funding during my years as a professor, I found myself sitting in the office of a senior administrator for one of the ten largest school districts in the United States. Each year, the administrator was faced with the task of making certain every child was assigned to a school, and that each school had sufficient capacity and resources to accommodate those children. The problem wasn't simple. In 1997, the school district operated over 260 schools with an enrollment of almost 200 thousand students. Due to population growth and geographic shifts in population, the district made extensive use of temporary, transportable classrooms known as T-buildings. When additional classroom space was needed on a particular campus, one of over 2100 T-buildings was loaded on a truck and delivered to that location.

During discussions with the administrator, I asked if an optimal assignment of students had ever been calculated for fixed school capacities; that is, rather than shifting buildings around to match the number of students in a particular school, reassigning the students to meet existing capacity limitations. Finding an assignment that minimizes the total distance students travel is one of the easiest optimization problems known. While political considerations make redrawing school district boundaries substantially more difficult than annually moving hundreds of T-buildings, the solution to the student assignment problem could be used to help the school district reduce student travel expenditures if it ever rethought district boundaries.

As I continued to describe some of the things science could do, she waved her hand to stop me. "You can do that?" she asked about the student assignment problem. "You can actually find the best way of assigning students to schools?" Yes, I said, and as I sat in the office with her, I could see her mind grasping something that had never occurred to her. There must be an almost infinite number of ways of assigning students to schools, but somewhere among all the possibilities there had to be an assignment that minimized student travel distance. A discussion followed in which I could see her realizing science's potential, but not quite sure what to make of it.

I wasn't surprised that the administrator didn't know how to actually solve her problems with science. What I found distressing was that she had no idea that science could be employed at all. Even though she worked with highly paid consultants, she'd never considered that there might be an assignment of students to schools that minimized buses, travel time, T-buildings, or anything else. I was very sympathetic to the practical and political realities of her job: the fact that she needed solutions and she needed to get those solutions implemented before the opening day of school. But I also had to wonder what additional burden taxpayers were being asked to shoulder because of what she didn't know. If costs could be reduced by 5, 10, or 15 percent just by knowing how to approach the problem, wasn't it worth a minimal investment in time and money?

To the administrator's credit, once she became aware of what was possible, it changed the way she thought about the problem. Earlier in my career, I was less successful with a proprietor named Ernie. Ernie ran two small but successful breakfast restaurants, one of which I frequented for many years. We liked to talk, and he once shared with me that he'd worked for a large oil company right after he'd graduated from college. "I told my wife they were going to make me president of the company or fire me," he said as he smiled and flipped a pancake.

On one occasion, Ernie asked what I did for a living. When my initial efforts to explain left him blank faced, I described my work in the context of a problem he might deal with. "Suppose," I said, "You have five packing facilities spread around the Midwest where you package breakfast meat—bacon, sausage, and ham. Thousands of different restaurants want deliveries of your meat on a weekly basis, and you want to find the least expensive way to supply them. What do you do?"

Ernie thought about it for a few seconds, then pronounced he knew the solution. While I didn't expect that he'd come up with the right answer, I was nonetheless encouraged. He was now thinking about the problem. How many trucks were required? What routes should they travel? What other information was needed to answer the question? With a little discussion, I was sure he'd quickly understand the complexities of the problem and realize that solving this and similar problems was sufficiently difficult to warrant making a living out of. I have never forgotten his answer: "Call Federal Express."

I tried to engage Ernie in a discussion about how Federal Express might go about solving the problem, but with no success. I persisted until I realized something that had never occurred to me. To Ernie, calling Federal Express *was* the solution to the problem. What was I making an issue out of? In the end, Ernie decided that I worked with computers. I happily agreed, and we left it at that. I was pleased when, ten years after opening his first restaurant, he retired comfortably before the age of 50.

Ernie certainly isn't alone in his approach to problem solving. Abstraction, generalization, and experimentation—the bread and butter of science and engineering—are difficult concepts. And as Ernie proved, there are situations such as running a small family business in which a limited understanding of these concepts is all that's necessary to successfully get by. For businesses of any size or complexity, however, science isn't a luxury, it's a necessity—especially when competitors are using science to their advantage.

At PROS, I've had the opportunity to interact with people in the pricing community at many different levels, from Fortune 100 CEOs and executives to managers and front line pricers. I am constantly impressed by the earnestness with which they all undertake their jobs. People involved with pricing have more war stories than most, and as a result, they tend to develop strong opinions about the right way to do things. Like traders on the floor of a stock exchange, pricers win and lose, day in and day out. In doing so, they develop a thick skin. How can they help it? "No" isn't an easy answer to take, whether it's read in falling sales figures or heard on the other end of the phone.

In one instance at PROS, I was participating in a sales engagement as we listened to senior representatives from a company that rented construction equipment. The corporate office was unhappy with the prices being offered to customers by its regional offices. While pricing guidelines were set at the corporate level, regional managers were given considerable latitude in the actual prices they charged. The corporate office believed the regional managers were taking advantage of this flexibility to offer lower prices than they needed to. Of course, the regional managers felt the corporate office was simply out of touch with the realities of the market. The story wasn't unique to this particular company.

The company rightly believed that pricing software would provide a competitive advantage by streamlining the pricing process and presenting data in a form that could be analyzed, thus helping resolve the conflict and finding the appropriate price. In attendance were assorted vice presidents and regional managers. At one point, one of the regional managers was moved to speak, succinctly making his position known to everyone in the room. "Look," he said. "I'm willing to give you a chance with this pricing stuff, but you've got to remember that my customers use f**k in every other sentence and start their negotiations by throwing down a price sheet from my competitor across the street."

I was impressed. First, he made his point that this wasn't a theoretical exercise. Second, he laid out two important attributes of the problem: his customers were a tough group, and competitors' prices couldn't be ignored. Third, he was willing to make a go of it. He was willing to give this "pricing stuff" a chance.

Further discussion helped explain why he was coming to the project with an open mind. Margins in many of the markets he was responsible for had fallen horrendously. Still, while the corporate office was pressing him to take a look at scientific pricing, it was also clear that he'd have fought like hell if he'd felt they were off the mark.

Even with an open mind, though, I knew it was going to take more than a fancy software system to convince him of anything. He needed results, and he needed results that he and the people working the counters could understand. He certainly wasn't ready for a group of PhDs to descend on him, journals in one hand and calculating devices in the other. At least, not right away.

Like the airlines many years ago, his company was just beginning its journey down the road of scientific pricing. Along the way, they were going to discover many things about their business and their pricing practices. In all likelihood, they would experience their biggest improvements as they came to embrace some of the most basic lessons of science: evaluating how individual retail locations were performing, which customers were really profitable, and the historical relationships between price and demand; and at the next level, how to forecast, what to expect from forecasts and how to deal with randomness, price optimization, and the importance of playing a consistent, winning strategy.

* * *

"Pure and simple" is how Vernon Lennon describes the mathematics he uses when training pricers. Then, as if he hadn't made his point clearly enough, he adds, "Simple, simple, simple."

Together with William Dudziak, Lennon is responsible for changing the pricing culture of BlueLinx to a more scientific focus. BlueLinx is the former distribution division of giant Georgia-Pacific, having once made up some 20 percent of the parent company's revenue. BlueLinx engages in selling building products predominantly to dealers and big box retailers. In 2005, plywood, lumber, roofing, insulation, and similar items earned BlueLinx revenues exceeding $5.5 billion. While the company deals with tens of thousands of individual SKUs, 80 percent of revenues come from less than 2,500 of them.

The nature of the business isn't one of fixed prices. The amount BlueLinx pays to get building products into its distribution warehouses changes on a transaction-by-transaction basis, as do the prices it charges its customers. A sales force of over 900 employees visits customers or works the phones. "What can you get me for two by six by twelve spruce?" comes a call from a customer. "Just a minute," responds the BlueLinx sales agent. Checking various computer screens, he determines a quote for the caller.

The important question is what takes place between the time the call arrives and the time the quote is made. Focus group sessions with sales agents uncovered concern over the cost information they were receiving—the amount BlueLinx paid to procure the goods they sold. The sales agents also admitted that with compensation based on volume, they tended to lean toward offering lower prices and making the sale.

Cost information was captured in gross margin calculations—sales price less average inventory cost. Pricing files that were based in part on cost information were maintained manually by a group of over 50 people, each of whom had his own way of calculating the "right" price. Should cost be calculated as a three month rolling average or as a result of personal knowledge about the markets? Should updates be made weekly, monthly, or at some other interval?

In reality, costs change with every order. If competitive market rates are rapidly rising, a sales agent can be selling sheets of plywood for $26 with an internally defined average inventory cost of $24 and appear to be making a great profit, even if the price to replace the plywood has jumped to $28. Not only was the company focused on cost-plus pricing, but it also wasn't using the right cost, not to mention that the sales force wasn't being incented to move with the market.

Another problem was that gross margin calculations only incorporated the cost of purchased goods and omitted other elements of the cost-to-serve, such as material handling and delivery. Sales agents were provided with rough estimates of transportation costs by being presented with concentric circles emanating from a warehouse. Within the closest circle, having a 50 mile radius, transportation cost one amount. This amount increased when a customer was located outside of the 50 mile circle but inside the 100 mile circle, and so forth. While the circles helped make salespeople aware of transportation costs, they didn't show up in final calculations, effectively leaving the importance of transportation costs to the discretion of the sales agent.

As part of a general pricing overhaul, Dudziak and Lennon refocused the company on contribution margin, replacing an ill-defined notion of cost with a very specific definition of current replacement cost. But this was only the beginning. As the result of a tremendous investment in computers and data, BlueLinx was better able to understand the cost of serving individual customers in much greater detail and the value of the company's services, and to price accordingly. Even such minor changes, however, created confusion, as I was to learn firsthand.

My efforts to sneak unnoticed into a sales training class were fruitless. Not only was the door situated at the front of the room, but my copious note taking left those in attendance wondering if they were under review. Standing in front of the room's white board was a replacement cost manager who was also one of a handful of "pricing pals" responsible for a combination of education, answering questions, and general cheerleading. The instructor and his fellow pricing pals were the people who were tasked with making the new pricing culture work. On this day, as on most, the topic was the new pricing system. The instructor was obviously enjoying his role teaching, sharing not only facts and insights, but humor and commentary as well. Eight men and two women were seated in the training session, with ages ranging from the early twenties to over 60.

The class was discussing a source of confusion that had come up. A sales agent had discovered that two customers with retail stores situated almost next door to one another were receiving two different price quotes from the system. With deliveries coming from the same warehouse and no appreciable difference in delivery location, the agent assumed the price to each store should be the same. The conclusion was that the system was broken.

The instructor began his explanation by reminding everyone in the room of the three components making up price: replacement cost, cost-to-serve, and science. Cost-to-serve, the instructor went on, incorporated accounts receivable. Customers with a good record of payment would naturally have a lower cost-to-serve. While this fact had been communicated to the sales force in an earlier training session, it needed repeating. As any teacher knows, homework isn't an effort to needlessly cause students grief. Real learning is only achieved through a combination of doing and questioning. The students had been to the lecture and done their homework, and now they were back questioning.

The discussion in the training room that day was lively and good natured, but with a sales force of 900, this isn't always the case. Drinking beers with some of the pricing pals and other advocates of the new way of pricing, I noticed how young the people at the table were. Was there a generational divide between those who found the changes easy to accept and those who didn't, I asked? The answer was a resounding yes, and the feelings poured out along with the beer.

I later talked with Dudziak about his thoughts on the subject. He agreed with the assessment, but hardly found it surprising. In Dudziak's mind, there were two contributing factors. First, the older someone is, the more set they become in their ways. When you do your job for a long time, you develop habits—good and bad—that become harder and harder to change. Second, technology has been around just long enough that today's students are comfortable with computers, spreadsheets, and other types of software. There's no exact age at which people on one side are computer literate and those on the other side aren't, and there are always exceptions. But as rule, there's a huge difference between someone who's close to retirement and someone who's fresh out of school.

Dudziak recounted a story from earlier in his life, before he came to Georgia-Pacific and BlueLinx. "One of the managers I knew had been trying to get his sales force to use laptops," he said. "One day, a salesman who'd been with the organization for many years and hadn't accepted the laptop announced in a sales meeting that he'd finally broken down and used it to write a memo. He then put the computer on his lap, pulled a sheet of paper from his bag, and demonstrated how he wrote the memo by hand using the laptop to support the paper." Asked what happened to the salesman, Dudziak told me, "He was asked to 'pursue other opportunities.'"

Dudziak was introduced to science early in his education. As far back as the mid-1980s when he was working on degrees in economics and civil engineering at Carnegie Mellon, every undergraduate was required to take a course in computer programming. Computer literacy was considered an important part of a college education for everyone. Dudziak never pursued a typical engineering career, however, instead working in a variety of sales, analysis, and financial positions in the building products industry. Along the way, he found time to complete a master's degree in management at Georgia Tech.

Dudziak has a soft spot in his heart for the academic world. He maintains a formal affiliation with his Atlanta alma mater and likes to work with students. Though he is anything but a geek, he's had his moments. In the course of a conversation on a recent first date, his companion politely asked what he did. Thirty minutes and a table full of napkins later, the spell was broken. A second date never materialized. His 15-year-old daughter understood perfectly. "Dad," she said, "You're *such* a dork."

Dudziak could only laugh, as he's known to do. Both he and Lennon like to think of themselves as the court jesters of the BlueLinx pricing effort. As seriously as they take their work, they know that a good dose of humor is the only way to deal with the many adversities they'll encounter.

Bringing up a new software system is tough, especially when scientific calculations are being performed on data from many different sources. Science has a tendency to surface bad data—a valuable result in and of itself, but not what people want to focus on when installing software. Dudziak and Lennon remember many long days working through various issues as they arose.

Without exception, everyone involved in the project pointed to managing change as the most challenging part of the implementation. The large number of sales agents alone presented obstacles. Every agent needed individual attention. As a result, the system was rolled out in phases based on geographical region. There were concept training sessions, system training sessions, and informal "popcorn and pricing" sessions. Once completed, they started all over again.

After the system went into production, sales agents were measured by how frequently they looked at the system's main pricing screen before making a quote. "Pinging the system," as this activity came to be known, was emphasized throughout training. Those who refused to use the system, showing few or no pings, were required to keep taking classes. Prior to a nationwide training event, humorous posters were displayed, each relating to golf and indirectly to Ping golf equipment. Within BlueLinx, the word "ping" is now synonymous with using the pricing system.

Counting pings is a strikingly simple way to measure people, but the results were equally striking. The average contribution margin per sale showed a significant positive correlation with how often agents pinged the system: sales agents who looked before they quoted offered the customers more appropriate prices.

Anecdotes abound about how the system was received. In one instance, Tony, a PROS scientist, recalls being approached by a sales agent who was unhappy because he'd just been "screamed at" by a customer over a "ridiculously high" price. As a result, the agent backed down and gave the customer the price he'd requested.

Talking to the sales agent and reviewing historical transactions, Tony discovered that the customer had been so adamant because he wanted the same price he'd received on his order five weeks earlier. Yet in the interim, the replacement cost to BlueLinx had increased substantially. The customer was basing his expectations on a reference price that didn't make sense given current market conditions, and the sales agent had just sold the item for less than it would take BlueLinx to replace it. The agent, being used to unit costs that remained unchanged for lengthy periods of time, couldn't understand why the system could possibly have quoted such a high price. Happily, when the sales agent understood what was going on, he became a huge advocate of the system. One down, 899 to go.

In another instance, a sales agent was on the phone looking to sell metal studs. Lennon remembers it in detail, as it was the first transaction made with the assistance of the new pricing and sales system. Taking into account the many different variables involved in its calculations, the PROS system quoted a price of $4.12 per stud. Using techniques similar to those employed before the system was in place, the agent arrived at a price of roughly $3.00. Giving the system the benefit of the doubt, but unwilling to entirely let go, the agent split the difference and offered a price of $3.50 for the first unit and $4.00 for the second. The customer happily took them both. Within the next few days, it became apparent that a run on steel in China was making itself felt around the world in the form of higher prices for manufacturing.

What really caught the attention of management, however, was that the system showed that a very large, national retailer had been receiving a price of $2.39 for the same studs under a special program—well below replacement cost. BlueLinx's national account representative was contacted, and discussions with the retailer led to a more equitable arrangement.

In yet another instance, a sales agent was working in a part of Texas that was perennially short on delivery trucks. By eleven o'clock in the morning, he had no shipping capacity for the following day. Anything sold thereafter would have to go out via a third party, significantly increasing the cost of delivery.

Described as a "rough and tough Texan," he was not yet convinced he was ready to adopt the pricing system, but decided to use it on a trial basis. For the rest of the day, he quoted what he considered a high price for one particular commodity. As the story was later related from the sales agent, two customers hung up on him, but another twelve took the price that was offered. And one of the customers who hung up called back and placed an order.

Anecdotes like these come as the result of tremendous effort, and at times, Dudziak and Lennon grew frustrated. Both recall a meeting with then-CEO Charles McElrea at a time when they felt the project wasn't progressing as rapidly as they'd hoped. After listening for a time, McElrea stopped them. His message: Sure there's a way to go, but look at how far we've come!

Dudziak and Lennon speak of McElrea's passion for pricing, a passion I observed for myself when I first met him. In Atlanta for a professional meeting, a senior scientist in my group and I met McElrea along with some ten or so senior executives from what was then the distribution arm of Georgia-Pacific. The meeting was held over lunch in an old, established, and very well-appointed restaurant in downtown Atlanta. It was everything I'd envisioned, from the finely dressed waiters to the elaborately framed pictures on the wall.

The discussion was pleasant, with McElrea wanting to learn more about PROS. A distinguished gentleman nearing retirement age, McElrea was gentle and soft spoken, but left no question as to who was in charge. As people introduced themselves, it struck me that their average tenure at Georgia-Pacific approached 30 years. Even the short-timer checked in with more than ten.

Somewhat to my surprise, throughout the discussion, it was evident that McElrea was doing more than learning about PROS. He was also introducing his team to the idea of scientific pricing. As I talked about examples of pricing and what we did, I was met with probing questions from many of those at the table. McElrea himself was questioned as his team politely challenged ideas that arose. This was a very early step on the road to scientific pricing.

While science was discussed in the most general terms, late in the lunch, McElrea stepped up with some questions about what mathematical techniques we employed, using explicit technical terms such as *linear programming*. As he spoke, McElrea's eyes lit up, and so did mine. Given the setting, we never had the opportunity to continue the discussion in any detail, but we'd made a connection. After lunch, my partner from PROS expressed the same reaction I had. It wasn't that we were excited to talk mathematics with the president of a $5 billion organization in a $25 billion corporation. Rather, we were excited because he understood what mathematics could do. With that level of executive understanding and the spark of enthusiasm that came with it, we knew that if given the chance to do a pricing project with the company, it couldn't help but be successful.

McElrea championed the project until his retirement in October 2005, when Stephen Macadam took over as CEO. Full of enthusiasm as he looks to the future, Macadam is focused on where BlueLinx will go next.

Macadam started his education with an undergraduate degree in mechanical engineering at the University of Kentucky, where he was introduced to the mathematical rigors required of the profession. He later went on to receive a master's degree in finance from Boston College and an MBA from Harvard. At Harvard, his achievements were recognized when he was named a Baker Scholar, an honor reserved for the top 5 percent of the graduating class.

Sitting with Macadam at lunch, I couldn't help but be impressed. Exuding energy, he's a leader, yet approachable in a way that many leaders aren't. And it's clear that he enjoys the business as much as he enjoys doing business.

OSB (oriented strand board, which I learned are boards made of wood flakes held together by resin) is less expensive than plywood because it can be made from small diameter trees using a less labor-intensive process. With today's resins, it's nearly as strong as plywood. Recent years have seen the demand for OSB skyrocket as it has replaced plywood in applications once dominated by the layered board. OSB is considered a commodity, a product with such common characteristics from producer to producer that no one who buys it really cares where it comes from.

Macadam was concerned about a product from one supplier that's constructed by placing a thin coating of reflective material on one side of OSB, serving as a radiant barrier. It turns out to be useful in hot southern climates, where the cost savings on air conditioning can be substantial.

"This new OSB product isn't a commodity, but the producer is pricing it like one," said Macadam. With such a great new product, he felt there was an opportunity to claim a price premium, at least in the near term.

The story was indicative of Macadam's thoughts about the building products industry. When I questioned him about how he saw the industry changing, his overall message was "slowly." Building products aren't like consumer electronics. Changes occur at a sedentary pace. Even if there's a great new product break-through, an occurrence that isn't common to begin with, it takes time for builders to alter their long developed habits. Reflecting the products, ways of doing business change at about the same pace.

Macadam can't stand the industry on its head overnight, but he's chipping away, and he's starting with his own company. We shared many ideas, some he's only pondering, some he's investigating even now. All had a science component, which wasn't unexpected given that this was my line of inquiry. One idea was that of a sales simulator. Pilots for airlines go through recurrent training. Why not sales agents?

I was intrigued by the idea. It would be simple enough to set up games in which a sales agent interacts with a scripted trainer. But Macadam was a step ahead of me, thinking of a computer program that would substitute for the trainer, automatically keeping score. He was thinking of a simulator, just like a pilot. Would it talk to the sales agent or print sentences on a computer screen? Perhaps there was value in letting the computer express verbal outrage as a means of making a point?

Whatever form a simulator might take, however, I knew there was science lurking in the background. After all, if you want to train sales agents how to act in sales situations, you need to make sure that the simulator is acting like customers would—not perfectly, of course, since this can't be achieved; but close enough to instill sales agents with the proper strategy for dealing with common occurrences.

System dynamics came up as another topic. Sales of building products are cyclical. Some of the cycles are predictable, such as the time of year. Others seem random, as buyers run up the price of an item through lively buying, then stop, causing the price to drop. The activity isn't entirely like watching the price of a stock as it rises and falls, since building products are constantly being consumed and replenished. Can these short-term cycles be understood and explained by feedback loops, or are they best modeled as random events, he wondered.

Riding back to the office, Macadam turned his attention to more immediate concerns. While the new system was giving the sales agents a much better mea-sure of cost-to-serve than anything they'd had previously, he was still concerned about accounting for the cost of full and partial shipments in more detail. Once a customer places an order for one item, the marginal cost of delivering a second item is almost nothing as long as there remains space in the truck. How should this be accounted for when selling and, more importantly, what did it mean in concrete terms for BlueLinx's profits? We both agreed that that there was the potential for everyone to win, with cost improvements coming from better efficiency. I loved the idea—driving efficiencies on the cost side by managing demand in the sales process. This was taking the concept of supply chain management to its logical conclusion.

In the end, I walked away with enormous respect for what BlueLinx was achieving. The basic formula was simple: believe in the power of data and commit to making it work for you. As Dudziak put it, "We realized that our value wasn't our trucks." Macadam went one step further, observing that in a world with so much technology at our disposal, "it's important to out-think your competitors."

To date, BlueLinx's scientific efforts aren't exceedingly sophisticated, but then again, they shouldn't be at this point in the company's development. As the implementation team quickly learned, the biggest challenge was making certain sales agents used the system and used it with confidence. With the focus on getting better information to the agents, early efforts with the pricing system have helped BlueLinx do away with bad business—transactions that lose money. With recommendations that aren't always intuitive, helping agents understand where the numbers come from is an important part of getting the most from the system.

Going forward, however, there remain tremendous opportunities for BlueLinx to leverage its investment. Only the mind of Steve Macadam and the team of people who work for him know exactly where BlueLinx is headed, but one thing is for certain: BlueLinx knows that a rational, thoughtful, informed business strategy is the only path to take. And science will be right along for the ride.

* * *

With pricing-related technology all around us—computerized inventory systems for tracking our purchases, machines that scan our groceries, Web sites that search the Internet for the lowest prices—it's natural to conclude that businesses are technologically savvy when it comes to the prices they set; that when we reach for a box of cookies or a bottle of aspirin, the number affixed to our purchase was carefully researched and calculated using all available information. Perhaps the owner of the independent restaurant around the corner simply uses his best guess, but certainly large corporations must invest considerable time and effort in the pricing process.

Yet with few exceptions, this simply isn't the case. The state of pricing remains distressingly seat-of-the-pants. Companies like BlueLinx are a rarity. More frequently, companies are in the same state as the rental equipment company. They know they can do a better job of pricing and they're looking for science to help them, but they need to overcome basic organizational and process issues first. What's frightening is the number of companies that, like the school administrator, are completely unaware of what science can achieve. But whichever example you choose, the reality is that pricing similar to what the airlines practice remains years in the future.

BlueLinx is fortunate. Thanks to visionary management, it has begun the transition to a scientific pricing organization earlier than most. More advanced science will come with time now that the process has begun, and BlueLinx is already in a much better position to out-think its competitors. The question is what will happen to organizations that don't begin the transition while their adversaries do.

CHAPTER 11

Pricing's Many Faces

The world of science is home to equations that are as famous as Harrison Ford is to the world of Hollywood. Many of the equations require advanced courses in mathematics to understand, but others derive fame from their very simplicity. Science is often associated with complexity, but in reality, scientists and mathematicians cherish simplicity. The highest praise one mathematician can offer another is captured in the word *elegant*. Elegant is used to describe the proof of a very difficult result in a creative and brief manner. Proofs that go on for hundreds of pages, even if wielding powerful mathematical machinery, are not considered elegant and do not receive the same praise as a five page proof of the same result. Of course, results with a proof that's too brief and that doesn't rely on some pillar of the mathematical edifice are at risk of being *obvious*, even if the proof took years to discover. Obvious cuts deeply into the ego, and is only surpassed by "there's an error in your logic." Many a mathematician has been cast into a state of despair when a senior colleague has declared a result obvious, and many pitchers of beer have been consumed debating whether or not a particular result is or isn't in the dreaded category.

For scientists, who rely on experimentation rather than the deductive reasoning of mathematicians, an equation's fame is measured by the territory it covers, the importance of what it explains, and how concise it is. Einstein's $E = mc^2$ is the most famous equation in history because of its association with the atomic bomb, but also because it succinctly expresses that very little mass m is required to create a lot of energy E. Had the relationship between m and E been any more complicated, it's doubtful we would remember Einstein's result in the form of an equation.

Not all equations are as widely recognized as $E = mc^2$, but many are equally famous within the communities they serve. *Little's Law* states that the average length of a waiting line is equal to the average rate at which people arrive multiplied by the average waiting time: $L = \lambda W$. What makes the result so remarkable is its generality. It doesn't matter exactly how people are arriving—one every minute, or in groups of ten every ten minutes. It also doesn't matter how people are being served—with one server or ten, randomly or not. Formulated by John Little, a pioneering operations researcher and now Institute Professor at the Massachusetts Institute of Technology, the result has been used to design waiting lines for everything from call-in service centers to routing information over the Internet.

The science of pricing has its own famous equation:

$$R = PQ$$

where R is revenue, P is price, and Q is quantity. (The equation can be cast in terms of profit by redefining P as price minus unit cost, and interpreting R as profit.) While it can't stand with the great equations of the physical sciences and mathematics—it's a definition, not a deep and unexpected insight into the world around us—the equation is simple, comprehensible, and fundamental to the science of pricing. Yet there is probably no equation in all of science that's so routinely ignored or used so inappropriately.

A look at the demand curve, often called the *price response function* in operational pricing settings, helps illuminate the misuse of the revenue equation $R = PQ$. Figure 11.1 depicts the typical downward sloping demand curve, showing how an increase in the price of widgets brings about a decrease in the number of widgets sold. The revenue at each price P is calculated by using the revenue

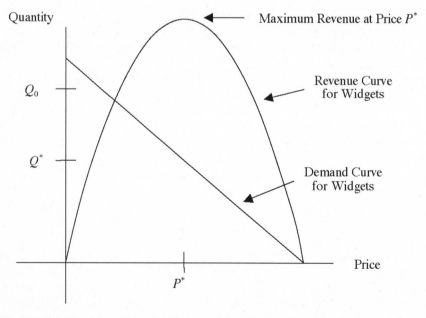

Figure 11.1 The demand curve and revenue curve for widgets. Maximum revenue is achieved at a price of P^* by selling Q^* widgets, leaving an unsold inventory of $Q_0 - Q^*$ widgets.

equation and the corresponding quantity Q from the demand curve. These revenue values are also plotted in Figure 11.1, creating a curve that rises from 0 when the price is 0 to a high point at price P^* before falling once again to 0 when the price is so high that no widgets are purchased. (The actual shape of the revenue curve is highly dependent upon the shape of the demand curve.)

A final feature of Figure 11.1 is the value Q_0 on the Quantity axis, representing the number of available widgets. Here we see what seems to be a paradox. Even though we've chosen a price P^* that generates maximum revenue by selling Q^* widgets, we still have $Q_0 - Q^*$ widgets left in inventory. These widgets must have some value. So how can revenue possibly be maximized?

The answer lies in the fact that to sell more than Q^* widgets requires dropping the price for everyone, and when this occurs, the net effect is to reduce overall revenue. Revenue management seeks to mitigate this problem by segmenting the market so that widgets can be sold at different prices. However, even when revenue management is employed, there are many times when the revenue maximizing decision is *not* to sell everything available in inventory.

In practice, however, most businesses unequivocally view unsold inventory as lost revenue, and, as a result, develop business processes geared toward moving volume. Sales departments receive commissions based on the number of units sold. When revenues are lower than expected, a common response is to drop prices and move more volume. In each case, organizations are focusing only on Q, ignoring the fact that R is the product of P and Q. The work done by BlueLinx can be summarized as an effort to strike the right balance between P and Q—a good first goal for any organization seeking to improve its pricing practices.

A second way in which the revenue equation is abused has to do with assumptions about the demand curve. The biggest assumption is that it's easy to find and that doing so is merely a matter of digging through data. Given enough of the right kind of data, demand curve estimation is, in fact, a straightforward process. The problem is that data is often lacking in a variety of ways.

The simplest yet most common deficiency arises when the data simply doesn't make sense. Anyone involved in data analysis will attest that a remarkable number of problems are encountered as a result of the way the data is defined or collected. Problems of this nature can lead to nonsensical scientific results until the data is *cleansed*.

Even when basic data cleansing is complete, there typically remain more fundamental data issues that must be addressed. Figure 11.2 shows the weekly sales data for a particular product over a period of nine months. With the product offered at only one price, there is no reasonable way to predict demand at other prices without additional information. Common solutions include price experimentation or deriving a demand curve from similar products. Figure 11.2 also shows that even at a single price, there is substantial variability from week to week. Much of the variability is attributable to unpredictable randomness, but over a nine month period, seasonality, holidays, special events, and changing factors in the market need to be accounted for—a process referred to as *normalization*. Still, for all the potential challenges, even relatively simple analyses yield important insights. Figure 11.3

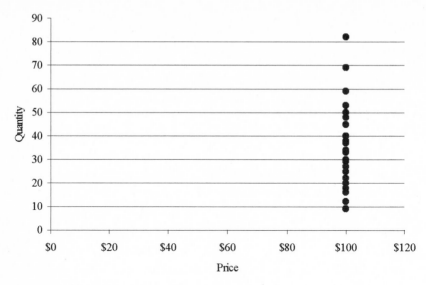

Figure 11.2 Weekly demand for a product as a function of price.

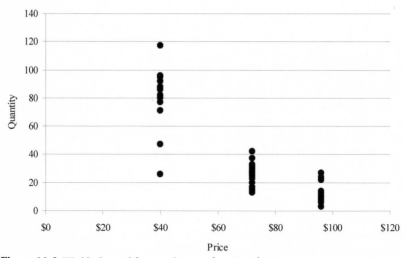

Figure 11.3 Weekly demand for a product as a function of price.

shows that average weekly volume drops significantly as price is raised from $40 to $70, and as a result, the revenue maximizing price is closer to the lower value than the higher value.

Frequently, a business has transaction data but it can't be used to estimate a demand curve. An example helps clarify this statement. Figure 11.4 shows sales data for widgets over a three week period. From this figure, it appears that consumers were more interested in purchasing widgets at the higher price of $5 than they were at $4. What the graph doesn't show, however, is that the item was offered

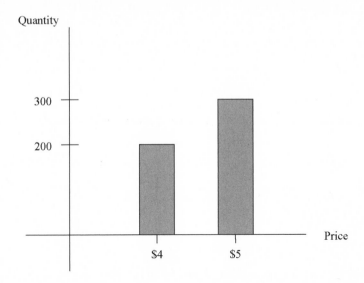

Figure 11.4 The total number of widgets sold over a 3 week period. A total of 200 were sold at a price of $4, while 300 were sold at $5.

at a price of $5 for two weeks and a price of $4 for only one week. When we plot average weekly sales as in Figure 11.5, the picture makes more sense.

The point of this example is that associated with every demand curve is an implicit length of time. When we speak of "the quantity sold at a given price," what we really mean is "the quantity sold at a given price for a specified length of time." Because explicit mention of time is rarely important in conceptual settings like courses in economic principles, it's often overlooked. But as the example demonstrates, time is vitally important. To estimate a demand curve, a price must be posted for a certain period of time and the number of purchases must be tallied at that price. Without this information, it isn't possible to construct a meaningful demand curve in any traditional sense. Without knowing the $5 price was offered for two weeks and the $4 price for only one, the total sales information isn't just useless, it's misleading.

Consider, then, the data displayed in Figure 11.6, where each bar represents a single transaction. Customers are now purchasing more than one widget at a time, and each purchase is subject to negotiation or to the terms of a previously negotiated contract. Suppliers commonly experience sales transactions of this type with distributors or retailers. With price on one axis and quantity on the other, it seems natural to use these transactions to estimate a demand curve, perhaps by fitting a line to the data. With no concept of a posted price, however, any demand curve created in this way is misleading in the same way as the data in Figure 11.4.

It's important to realize that the problem is that the transaction data doesn't support estimating a demand curve even though a demand curve may exist. It's

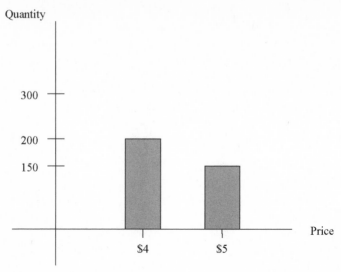

Figure 11.5 Accounting for the fact that widgets were priced at $4 for 1 week and $5 for 2 weeks, an average of 200 widgets per week were sold at a price of $4, while an average of 150 widgets per week were sold at $5.

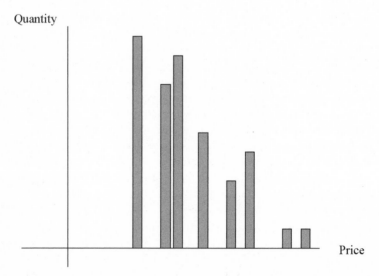

Figure 11.6 Each bar represents the number of widgets sold in a single transaction. It is wrong to estimate a demand curve by fitting a line to the observed data.

perfectly reasonable to imagine that if a supplier sold surgical gloves by posting a fixed price, then there's some quantity he should expect to sell at each price. The information contained in the transaction data of Figure 11.6 simply doesn't allow him to estimate this quantity. Surprisingly, many companies are awash in transaction data, yet still can't generate a traditional demand curve. "Water, water, everywhere," wrote seventeenth century poet Samuel Taylor Coleridge in the *Rime of the Ancient Mariner*. "[But not a] drop to drink."

A third way in which the revenue equation is abused is by assuming a demand curve exists when it clearly doesn't. The issue isn't one of faulty estimation or lack of appropriate data, but rather that there isn't a demand curve to estimate.

A good example is the case of two people negotiating the price of a unique antique vase. Would a seller ever consider that at a price of $1 thousand she could sell 42 vases, while at a price of $10 thousand she could only sell five? A demand curve simply doesn't capture the essence of the situation. A more common business problem is the negotiation of large contracts with many terms. Once again, attempting to force demand curves into the discussion is a mistake. Yet we are so familiar with the basic economic concept of a demand curve that it's difficult to break free of it. The lack of a demand curve doesn't invalidate the revenue equation. $R = PQ$, always and everywhere. But how to use this equation needs to be rethought.

* * *

Louise de Marillac was a woman of uncommon generosity. Born into the upper echelons of French society in 1591, she considered pursuing a religious vocation but instead was married at the age of 22. Her husband, a devout and honorable man, supported her personal efforts to minister to the poor, even as these efforts impinged upon on the sensibilities of those in their social circle.

When Vincent de Paul first met Louise, he was seeking to organize well-to-do women to serve the poor. With her piety, sense of conviction, and ability to brave the appalling poverty faced by those she attended to, Louise proved to be exactly what de Paul was looking for. Together they would form the Daughters of Charity of Saint Vincent de Paul, an effort for which Louise would ultimately be granted sainthood in 1934, and be named Patroness of Social Workers in 1960. Traveling throughout France, Louise helped the poor by establishing and operating hospitals and orphanages. The Daughters of Charity still boasts a religious community of some 27 thousand women scattered throughout the world serving the sick, orphaned, and poor.

The Seton Healthcare Network is located in central Texas in and around the state capital of Austin. Named after Saint Elizabeth Ann Seton, who was responsible for the growth of the Daughters of Charity in the United States in the mid-nineteenth century, it is part of the Ascension Health system—the largest nonprofit health organization in the United States, employing over 100 thousand associates.

Seton continues the tradition established four centuries earlier by Saints Louise and Vincent. Until recently, daughters would actively engage in many hands-on caregiving activities. But with the graying religious population and so few young people willing to take vows, most of the remaining daughters affiliated with Seton hold a small number of administrative positions.

Numbers aside, Seton's leadership and its underlying organizational philosophy are clearly those of the Daughters of Charity. Pictures of religious figures adorn the office walls, many showing women tending to the impoverished. "Our mission inspires us to care for and improve the health of those we serve with a special concern for the sick and the poor," reads the organization's mission statement, and activities reaffirm that Seton lives what it professes. In 1997, the Seton Healthcare Network led the way in forming the Indigent Care Collaboration, an alliance of local health-care organizations, volunteer clinics, and government agencies whose purpose is to serve the needs of the medically destitute. The Children's Hospital of Austin, part of the Seton network, developed a unique program with the Austin Independent School District to ensure that none of the 90 thousand public school students go without basic health-care services, and that uninsured children are connected with programs for which they are eligible. Through a variety of programs, including Seton Care Plus and Disease Management for Patients with Chronic Conditions, the network has reduced by half the number of emergency care and inpatient visits to its hospitals through preventive medical efforts.

There is a cost for these and other programs. In fiscal year 2004, Seton spent $85 million on charity and community benefit programs. With losses attributable to Medicare and Medicaid patients of $107 million, and $25 million in unreimbursed expenses (primarily from patients who have no insurance and can't afford to pay, yet don't qualify for government programs), uncompensated care of all types came to a mammoth $217 million. Net revenues during the same period were approximately $900 million. Like it or not, serving the medical needs of the poor is a business. "No margin, no mission" is an unwritten part of the mission statement understood by the professional health-care administrators responsible for the financial health of the Seton network.

John Evler is one of these administrators. Senior vice president of insurance services, he is responsible for all commercial insurance pricing, contract negotiations, and contract management. All totaled, over 70 percent of the Seton network's $900 million in net revenue fall under his purview. A strong yet unpretentious presence, sport coats and collarless shirts complement a hair-free head. His office is reassuringly plain, as are all the offices in the utilitarian building located just north of the Texas capitol building. Describing himself as a bit out of the ordinary for the health-care industry, Evler is scientifically savvy. Sharp and keen-witted, he has a penchant for storytelling and an obvious love of public speaking. Over his 25 year career in health care, he has worked in product development, marketing, sales, and physician practice development. Apart from a stint with an insurance company, Evler has always worked for independent, nonprofit health-care providers.

Evler's willingness to try new things is evident from his history. On one occasion while working in marketing for a health-care system in Orlando, Evler put together a combined print and television advertising campaign to raise public awareness of the hospital he worked for at the time. In the television ad, a businessman enters an elevator looking stressed. The sound of a normal heart beats in the background as the door closes. When the elevator door opens again, the sound of the heartbeat is now faster and more erratic, and the man's face shows obvious pain. Again, the door closes. As viewers watch the door open for a third time, they do not see the man, but stylized pictures of two heart monitors. On one monitor is the picture of a normal heartbeat, with the name of the hospital and the caption "live." On the other is a flat line, with the caption "die." While there is no specific mention of any other hospital, the ad caused such a stir in the tightly knit, local health-care community that it was pulled before the scheduled campaign was complete. It would later go on to win an award from the community of advertisers.

Evler understands the many idiosyncrasies of marketing and pricing health-care services. Health care in the United States revolves around service providers, with a distinction typically made between physicians and facilities. Hospitals are the most well-known service facilities, providing beds, operating rooms, specialized treatment wards, nursing support, and a host of other services. Conceptually, non-profit hospitals, which make up the vast majority of hospitals in the United States, determine "fair and reasonable" charges for what they provide with an expectation that their costs will be covered. In keeping with the nonprofit mind-set, hospitals tend to speak a language of services, charges, and reimbursements, not products and prices. Accounting is strongly represented in health-care education and among practicing health-care executives. Of course, if a hospital can't offer "charges" that are competitive in the market, it won't see enough patients to cover its expenses and will be forced to cease operations.

Very few people pay for health-care services directly. Instead, almost all services are paid for either by state and federal programs or by private commercial insurers (private charitable efforts make up a small but important component as well). A typical hospital might see 40 to 50 percent of its revenues come from the federal programs Medicare and Medicaid and another 40 to 50 percent from commercial insurers, though this can vary widely. Government programs are funded by taxes, while commercial insurance is paid for by premiums that insurance companies receive from the insured. Hospitals can have some impact on the payment they receive from government programs through lobbying efforts, but for the most part, the terms of payment are taken as given. Making matters worse, the accounting practices of virtually all hospitals show government-paid services occurring at a loss. To generate a positive margin, hospitals must therefore look to their commercial insurance contracts.

Commercial health insurance is complicated by the fact that most insured individuals don't contract directly with an insurance company but are covered through a sponsor, normally, their employer. The typical chain of payment for a working individual with health insurance is thus: (a) employee pays employer directly, indirectly, or both, (b) employer supplements employee contribution (often substantially) and makes payment to the commercial insurer, (c) commercial insurer pays health-care service provider for services rendered.

Price negotiations are an important element of the agreement between the employer and commercial insurer and the agreement between the commercial insurer and the health-care service provider. Commercial insurers are typically for-profit entities subject to highly competitive market forces. It is therefore in their best interest to negotiate advantageous terms (low prices) with health-care service providers so that they are competitively priced when their sales representatives get a call from an employer or sponsor.

For hospitals, the key source of revenue over which they have control is the payment they receive from commercial insurers, and this is where they focus their most attention. The contracts between these two entities establish the price insurance companies will pay when a covered patient shows up at the hospital door needing treatment. Set prices high enough and there will be money for charitable work and new equipment. Set prices too low and the hospital will be scrambling to pay its bills. A typical hospital may hold anywhere from a few dozen to a few hundred commercial insurance contracts, but will generate the vast amount of its revenue from no more than a handful of large contracts. Not surprisingly, the effort focused on a contract negotiation is proportional to the revenue the contract is likely to generate.

Commercial insurance contracts are Evler's domain. When I met with him, however, it wasn't immediately clear that contracts and pricing were going to be the topic of discussion. Instead, the conversation focused on safety. In the context of our discussion, safety embodied efforts to protect patients from avoidable illness or injury while at the hospital. Ventilator Associated Pneumonia, or VAP, can be a pervasive problem in hospitals. As the name of the condition intimates, the activity of wearing a ventilator for a prolonged period of time can lead to pneumonia that, once contracted, can cost up to $45 thousand per case to treat. Preventive measures that include periodically elevating the patient's head and weaning the patient from the ventilator as soon as possible can reduce the likelihood of VAP. Central line infection is a similar problem, as patients with stents inserted in a vein develop ailments that may cost as much as $25 thousand to treat, while good preventive treatment may cost as little as $10. Seton recently spent nearly $2 million on new bedding, in part to reduce the incidence of pressure ulcers that arise when a patient is confined to lie on a bed for an extended period of time.

I soon learned that Evler was concerned with efforts to improve safety because they impacted revenue. When a patient contracts VAP or any other ailment, the hospital gets paid to treat it. "Health care is one of the few industries where we get paid for our mistakes," said Evler. With fewer cases of VAP, the hospital could be looking at a multi-million dollar dip in revenue.

To be sure, Evler is all for safety improvements for obvious reasons, but he nonetheless must pay attention to revenue. By increasing the safety of patients and reducing cost for the insurance companies, Evler believed the hospital was entitled to a share in the financial benefit. But he also realized that commercial insurers were more likely to say "thank you very much for your efforts at improving patient safety" than to write a check.

To address his revenue concerns, Evler realized that safety represented an important source of market differentiation between hospitals in the minds of consumers. Patients care about safety. If the patient population knew about Seton's focus on safety, wouldn't they demand that their insurer offer Seton as an option? Wouldn't that put Evler in a better negotiating position? "We're always looking for the high ground to fight from," said Evler. "If we're safer, how much is that worth to you?"

As a seasoned negotiator, Evler knows how to look for the high ground. In one instance, he realized that one of Seton's competitors couldn't offer uniform prices or service because hospitals in the competing network operated independently of one another. Seton, on the other hand, operated under a centralized model, allowing a much more uniform degree of care. "I don't know if you're old enough to remember," said Evler, "but there was a time when you chose your roadside restaurant by the number of cars in the parking lot. Now we go to McDonalds or Burger King because we demand a uniform product." To make his point, I was then treated to an elaborate and entertaining story about the number of mushrooms on a favorite hamburger.

Evler was also grappling with how to price radiation treatment for brain tumors using a newly acquired piece of medical equipment known as a CyberKnife. The latest advance in radiation treatment, the CyberKnife uses robotics and advanced computer programs to reposition the radiation beam with great accuracy and without the need for immobilizing the patient. Its technological predecessor, the GammaKnife, requires intrusive immobilization and considerable skills on the part of the radiation specialist.

Seton never purchased a GammaKnife, but the competing network did and had yet to purchase a CyberKnife. Both the CyberKnife and the GammaKnife allow patients to return to their normal daily activities with almost no time in the hospital, but the experience is far more traumatic with the GammaKnife, requiring the placement of a mechanical apparatus on the patient's head to help position the radiation beam. While Seton purchased the CyberKnife to improve patient care, having spent many millions of dollars on the new piece of equipment, Evler was looking for ways to translate the purchase into a better competitive position and more favorable contracts with insurance companies.

Anyone who has ever negotiated a contract knows that it's important to enter the negotiation with a position that's as strong as possible. Reputation, quality, stability, and even more subtle issues play into the final terms of a contract. However, where Evler and his team go far beyond other hospitals is in their scientific analysis of contract terms.

Pricing health care is a remarkably complex task. There are many different ways a hospital can price its services, and over the years, most have been tried. At one end of the spectrum, a hospital can establish a very detailed list of charges, with a price for everything from the use of a heart monitor to administering aspirin. In fact, hospitals do keep a very detailed charge list of this type that helps for accounting purposes. Using this list, the charges incurred during a patient's stay can be tallied to determine what the hospital is owed. When commercial insurers and hospitals agree to an arrangement in which payment is made for detailed charges, negotiations focus on the discounts that will be offered, and agreements of this type are referred to as *discount-off-charges* contracts.

As these contracts have evolved, list charges and net charges have grown ever further apart as bigger discounts are offered from year to year. Discounts in the range of 50 to 60 percent are now common. This level of discounting may seem outlandish, but similar discounts are common in many industries and are easily observed in the retail market. Just because an item of clothing is marked down 50 percent doesn't mean the store isn't making a profit. With hospitals offering such large discounts, list charges rarely reflect actual cost. But since they were once intended to do just that, hospitals continue to routinely report charges rather than net charges. Thus, $2 billion in charges—the monetary proxy for the amount of work a hospital has performed—may correspond to only $1 billion in actual money received.

Discount-off-charges contracts have the consequence that hospitals and other service providers are given no financial incentive to keep costs down. Longer stays and more expensive procedures generate more hospital revenue. This is not to say that anyone in the medical community would knowingly promote unnecessary treatment, but anything that falls in the gray area is likely to get resolved by opting for further treatment.

At the other end of the spectrum are *capitated* contracts, which place the entire responsibility for cost of treatment on hospitals. With capitated contracts, commercial insurers pay a flat amount to the hospital for each insured person. This amount is paid whether the patient needs extremely expensive treatment or no treatment at all. The money a hospital receives under such a contract is essentially fixed. If the cost of treating all patients over the year is less than this amount, the hospital shows a positive margin. If an abnormal number of expensive cases comes through and drives up the cost of treatment for the insured population, the hospital loses money. In effect, insurance companies, whose primary purpose is to assume risk, pass the risk on to the hospital. While capitated contracts had a brief period of popularity as commercial insurers gained the upper hand in contract negotiations, they are now found only infrequently.

Per diem and *case rate* contracts offer two intermediate forms of charging for services. With per diem contracts, hospital services are bundled and a fixed fee is charged. For example, a one day stay in an acute care facility within a hospital may be charged at $1,800 per day independent of the specific nursing activities that occur during that period. Stays in an intensive care facility might cost $2,200 per day. In this way, hospitals are incented to exercise a reasonable degree of restraint when staffing and managing facilities, but unlike capitated contracts, they are paid for each and every bundle of services provided. If a patient needs to stay three days instead of two, the hospital will receive more money for the additional day.

Case rate contracts structure payment around the type of ailment afflicting a patient. Thus, for example, once a patient is diagnosed as having viral meningitis, the terms of a case rate contract might dictate payment of $2500 to the hospital. The advantage of such contracts is that they provide an incentive for hospitals to perform cost-effective services without shifting the entire burden of risk away from the commercial insurers. Insurers bear the risk related to the number of patients requiring treatment, while hospitals bear the risk associated with the cost of treating individual patients—something they have a degree of control over.

Instituting a case rate contract requires a classification scheme that allows physician diagnoses to be assigned to something billable. If a patient is diagnosed as having a "nontraumatic stupor and coma," hospital accounting needs to know there's something corresponding to this diagnosis in the contract, and then bill this amount. With so many potential ways to classify diagnoses, case rate contracts represent a potential nightmare. Fortunately, Robert Fetter addressed this problem in groundbreaking work at Yale University in the 1980s. Introducing a classification scheme bearing the simple but descriptive name *diagnosis related groups*, or simply DRGs, to most health-care practitioners, Fetter was responsible for changing the way the federal government reimbursed hospitals. Originally using a cost-based reimbursement approach similar to a discount-off-charges contract, Medicare shifted to a case rate structure with DRGs defining the patient classification scheme beginning in 1983. For his efforts, Fetter won the Edelman Prize, claiming savings in Medicare payments to hospitals of over $50 billion by 1990.

All basic contract types may incorporate special provisions. Some are relatively common, such as tiered reimbursement structures. For example, a *stop loss* on a discount-off-charges contract requires a commercial insurer to pay increasingly higher rates once discounted charges exceed a specified threshold. *Carve outs* are reimbursement methods that either party wants to single out for special treatment.

If anything should be clear from this discussion, it's that figuring out what a hospital actually sells and how it should be paid are remarkably complex questions. When we purchase a toothbrush we know what we're getting. When we need heart bypass surgery, we receive dozens, even hundreds, of services. Even worse, your bypass surgery may wind up costing far more than mine, even if our conditions appeared identical when we entered the operating room.

The pricing problem is so complex that no one's really sure what to do about it. Solutions are as varied as the many contract negotiations that take place, but there are recurrent themes. Straight line increases are common, whereby the basic contract structure established in the past remains in place, but prices, discounts, or both are increased by a fixed percentage across the board. New stop losses or carve outs may be incorporated, often the result of specific problem cases encountered in the prior year. A small set of strategic goals may come into play, for instance, if a member of the hospital's board of directors is concerned that a particular commercial insurer has been making noise about not renewing.

Asked if he has any humorous anecdotes about contract pricing or price negotiations, the normally jovial Evler simply shakes his head. "Any stories I can think
of are pathetic," he says. "How do you have a multibillion dollar industry that
operates like this?" Evler isn't exaggerating when he refers to the size of the problem. In 2004, health-care expenditures in the United States totaled $1.9 trillion,
or approximately $6280 for every person. Nearly one-third of this money was
paid to hospitals, meaning that payment to hospitals alone represented over 5
percent of the total gross domestic product of the entire United States. Contracts
between hospitals and commercial insurers aren't just big business, they're huge
business. Those who come to terms with the complexity of contract pricing stand
to reap similarly huge benefits.

One of the most basic activities a hospital can undertake when evaluating a
contract is to calculate the revenue it is expected to generate. In many cases, the
simplest way to achieve this is to take the patient cases covered by the contract in
the prior year and calculate the charges and margin the new contract would generate in the coming year under the newly proposed terms. In essence, these
patients multiplied by these contract terms equals this much revenue. Yet even
this conceptually simple computation is wrought with challenges brought about
by the many different contract structures and the many different services hospitals provide. Among the hospitals that attempt to perform this calculation, the
process often takes weeks and a staff of people drawing data from many disparate
sources before setting up spreadsheets to make the calculation for a single contract proposal. Many hospitals simply don't bother.

Seton recognized the importance of this activity and undertook the implementation of a contract modeling system to streamline the process and reduce
potential errors. Along the way, the hospital ran into its own set of challenges.
Dave Cripe, who works for Evler, recalls that something as simple as counting
patient encounters proved to be a challenge. No matter how carefully the calculations were checked, the number of encounters always proved to be higher than
the number calculated by the hospital's finance department. After much tribula-
tion, it was discovered that the finance department didn't treat newborns as
encounters, whereas the contract modeling system did.

The ability to quickly and accurately calculate a contract's potential revenue is
extremely important, but it's just the beginning for scientific contract modeling.
Another opportunity is to estimate contract revenue using forecasted demand,
not last year's patient encounters. If the treatment of ear infections is falling, perhaps because patients are being attracted to new suburban clinics, this is invalu-
able information. Going beyond what happened last year and focusing on what
is expected to happen next year provides a competitive advantage. A hospital
negotiator can target price increases on activities with growing demand while giving concessions in areas where demand is falling. If health-care services were as
simple as boxes of printer paper, similar methods wouldn't work. You want to buy
at a low price, I want to sell at a high price, and we both know the price when
we're done. Ironically, it's the complicated nature of health-care services that provides scientifically savvy negotiators with tremendous opportunity. Realizing this

opportunity simply requires making a commitment to deal with the data and the numbers, something many in the health-care industry seem unwilling to do.

Moving further into the scientific realm, optimization can be used to propose contract terms that meet criteria established by the negotiator. For example, if a hospital negotiator wants to improve overall margin in the coming year by 3 percent, an optimization model can be set up that will return terms that achieve this goal. Of course, without any further information, the computer is likely to respond with many possible solutions—there are many ways to raise prices! However, by incorporating constraints the commercial insurer is sensitive to— for example, a desire to keep a lid on emergency room costs—the goals of both parties can usually be achieved.

Another component that can be incorporated in optimization models is risk. At one hospital, an analysis of a contract with one commercial insurer showed a total of 900 encounters during the fiscal year. The contract generated a positive margin, but further analysis showed that only 88 of the 800 encounters actually generated payments that exceeded costs. Even more, a very small number of these 88 cases generated most of the margin. Put another way, the hospital was depending on a few catastrophic patient encounters for the contract to make money. Since the variability of these rare events was high, the risk of loss was also quite high. Optimization can recommend prices that minimize this type of risk.

At this extreme end of the scientific spectrum, health-care contracts become portfolios of assets. Each contract term represents a potential source of revenue for a hospital, with the actual revenue dependent on random patient encounters. Much as mutual fund managers balance risk and return as they put together a portfolio of stocks, hospital contract negotiators create a portfolio with inherent risk and return with the contracts they negotiate. Mutual fund managers employ highly sophisticated techniques when balancing their portfolios. Most health-care providers have yet to think of their contracts as a portfolio of assets.

The key difference between a portfolio of stocks and a portfolio of health-care contracts is that stocks are purchased and sold in an open market, whereas contract terms are negotiated. Because of this, there will always be important intangibles in the pricing process. The art of contract negotiation is structuring the negotiation to achieve desired goals. The science of contract negotiation is understanding what those goals should be. The more complex the terms, the more science can bring to the table. For negotiators, science isn't always the most enthralling aspect of a negotiation, but when the stakes are huge, science simply can't be ignored.

* * *

The price tag is alive and well. Whether it's stuck on the side of a soup can or listed next to a picture of a handbag on a Web page, the price tag will always be an important cornerstone of modern commerce. Traditional scientific pricing was designed for the world of the price tag, where sellers post prices and consumers take it or leave it. It's therefore not surprising that scientific pricing has flourished in the travel and retail industries.

Many of the world's important business transactions, however, don't involve a posted price. They may involve negotiations that range from a five minute phone conversation to discussions lasting months or years. Thousands of items or just a handful may be under consideration. Terms may involve payment for an established number of items, or the right to make future purchases at an agreed upon rate. But whatever the details, all businesses ultimately generate revenue according to the same equation: $R = PQ$.

Leading scientists have made important contributions to understanding the many diverse pricing problems, but for those that fall outside of the traditional domain of price tags, much work remains to be done. An important part of the work is simply breaking free of demand curves when they either can't be estimated or don't exist. The demand curve is such a deeply held belief that we seem to fall back upon it whenever "science" and "pricing" are used in the same sentence. Rather than continue hammering square pegs into round holes because that's all we've got, we need to step back and fashion some round pegs.

Fashioning round pegs, however, is only the first step. We then need to use them. Seton embraced a new way of looking at the world, but it required vision and commitment. Though Evler and his team are using powerful mathematical tools to achieve their mission, demand curves are nowhere to be found.

Alternatives to the demand curve aren't limited to occasions in which they don't exist. Inspired by the Internet, experimentation is one of the most promising new approaches to pricing. Rather than searching through historical transactions for hidden price relationships, these methods take a more blunt approach to determining what the market wants: try it, and if it works, keep doing it. The experimental approach is especially well suited to Internet sales and to compound pricing problems—problems involving the sale of a family of ten television sets, not just a single model. Science can be used to generate a pricing scenario for the family of items being sold, but the scientific focus is on evaluating whether scenario A is better or worse than scenario B. And experimentation isn't limited to price. Elements of the sales experience, such as where the pictures of the different models of television sets are placed on the Web page, can also be evaluated.

The experimental approach allows two or more vastly different pricing scenarios to be compared with one another. What it doesn't do is attempt to answer many detailed questions, like how a price increase of $20 in one model of a television set impacts demand for the other nine. Questions of this nature aren't just asking about the demand curve for a single model of television set, but rather the joint demand curve for all ten—an elusive prey at best.

One of the signs of a mature discipline is a structure that, in one form or another, most people agree upon. Modern mathematics is built on a foundation of calculus and analysis, then broken into fields of specialization such as partial differential equations, algebra, topology, and many others. Economists study micro- and macroeconomics. Logic is a requirement for the study of philosophy, and there are important schools of thought that every well-educated philosopher is presumed to know about. Some disciplines have such well-defined structures that we forget they are the product of centuries of human thought. Today's chemistry, with its periodic table of the elements, isn't anything like chemistry of the past.

The sign of a young and growing discipline is that it's in the process of discovering what it's all about. Pricing isn't new, but the focus on using detailed information about sales transactions to improve pricing is, and the reason is the power of today's science and computers.

Traditional revenue management and retail pricing have developed a relatively consistent science, but structured fields of pricing specialization haven't emerged yet. To some extent, the structure is evolving industry-by-industry, business problem–by–business problem. But the broader structure, the set of mathematical abstractions that transcend a particular business problem and define when two problems are fundamentally similar, have yet to become clear. The issue isn't one of developing general pricing principles, principles like avoiding cost-plus pricing or not allowing a competitor to control prices in the market. Rather, the issue is one of *science*—understanding how to use hard data in ways that drive higher revenues. For pricing science, the next decade promises to be an exciting time.

Notes

- Historical information on Louise de Marillac was taken from http://www.saint patrickdc.org/ss/0315.htm#loui.
- Information on the Seton Healthcare Network can be found at http://www .seton.net/about_seton. Charity and uncompensated care figures are taken from the Seton charity care and community benefit report (2004). For more information on Ascension Health, see http://www.ascensionhealth.org/about/statistics .asp.
- Statistics on national health-care expenditures were taken from the Web site of the U.S. Department of Health and Human Services Centers for Medicare and Medicaid Services, http://www.cms.hhs.gov/NationalHealthExpendData/down loads/highlights.pdf.

CHAPTER 12

The Coming Revolution

Nicolaus Copernicus (1473–1543) was many things—diplomat, soldier, economist, mathematician—but he is best known for his work as an astronomer, and in particular, for proposing a heliocentric theory of the solar system. Geocentrism, which places the earth at the center rather than the sun, had long been the predominant viewpoint for the simple reason that it was obvious. The earth wasn't moving.

The history of heliocentrism spans generations and is one of the greatest stories in the evolution of human thought. Galileo Galilei's (1564–1642) battles with the Catholic Church are legendary. Sir Isaac Newton's (1643–1727) laws of motion, which elegantly describe planetary movements about the sun, were intimately tied to the acceptance of heliocentrism.

What Copernicus started was more than a centuries-long search to determine what goes around what. The Copernican Revolution doesn't just refer to changing theories about the solar system. It refers to humankind's realization that we're only a part of things, not the center of attention. The metaphor is perfect, moving the earth from the center of the universe to somewhere less important. But as years of Church opposition demonstrated, displacing the earth wasn't metaphorical, it was real. God created the universe *for* humankind—the skies above and the seas below. It was simply unimaginable that we could be anywhere *but* the center of the universe.

Seeking to restore us to our rightful place, businesses are now focused on developing customer-centric shopping experiences. The metaphor is not lost on proponents of the movement, who describe the customer-centric "worldview" as creating a business that "revolves around the customer." Customer relationship management (CRM) software remains one of the most popular enterprise applications, reminding us that the issue is more than customer service, it's the customer relationship.

As the market for CRM software evolves, however, businesses are realizing that the dollar value of a customer relationship is difficult to measure. The early adage that it takes five to ten times as much money to find a new customer as it does to retain an existing one is no longer adequate justification for spending tens of millions of dollars on software. A good relationship is a nice start, but it needs to generate money.

One of the great challenges facing pricing, particularly in retail applications and the travel industries, is how to incorporate it in the customer-centric framework. At first the thought seems shocking, bringing to mind images of looking at Evan and Amy, then telling Evan he needs to pay $3 for a loaf of bread and telling Amy she only needs to pay $2. The very thought affronts our sense of justice so deeply that our immediate reaction is to refuse even discussing the possibility, and it raises the question of whether pricing this way is even legal.

It's important to recognize that setting a price using customer information is different from simply charging different customers different prices. When it was first introduced by the airline industry, revenue management was fundamentally *about* charging different customers different prices. But this was achieved by defining products in the way of fare classes and then either making

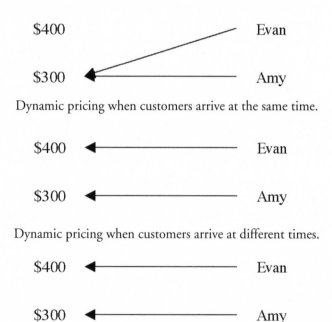

$400 — Evan
$300 ← Amy
Dynamic pricing when customers arrive at the same time.

$400 ← Evan
$300 ← Amy
Dynamic pricing when customers arrive at different times.

$400 ← Evan
$300 ← Amy
Customer-centric pricing when customers arrive at the same or different times.

Figure 12.1 With dynamic pricing, when Evan and Amy arrive at the same time, they see identical prices, but if they arrive at different times, they may see different prices. Customer-centric pricing provides for the possibility that Evan and Amy both arrive at the same time, but based on their individual characteristics, they see different prices.

Traditional revenue management operates similarly, but uses differentiated products to charge different prices. If, for example, the $300 and $400 products represent airline fare classes, Evan and Amy might arrive at the same time and observe that both are available, but because Evan wants a fully refundable ticket and Amy does not, he pays the higher fare. However, the airline has not "looked to see that it's Evan" at any time during the transaction.

them available or not. At no time did airlines actually look at the customer at the time of sale and say, "Oh, it's Evan, I'm only going to make my most expensive fare class available to him." Customer-centric pricing considers the possibility of taking this final, logical step.

Figure 12.1 makes this point graphically for a product that can be offered at one of two different prices. Science uses historical information on how the product has sold at the different price points to determine what price makes sense at any given point in time. This price may be varied over time, but once the price is established, any customer who shows up can purchase the product at that price. Science and practice have historically focused on this type of dynamic pricing. Customer-centric pricing adds an additional layer of complexity as knowledge about the customer is used to further refine what price is offered. In the extreme, the product pricing layer can be done away with altogether.

Frightening as the thought of customer-centric pricing might be, there are many situations in which it's not only acceptable, but socially desirable. It costs more for a plumber to cross town to fix a sink than to fix a sink next door, and no one would fault the plumber if he charged more in the former case. Senior citizens are routinely offered discounts for everything from golf to meals in restaurants. Insurance companies have long used customer-centric pricing. Homes with security systems are less likely to be robbed and therefore receive discounts. Flame-retardant building materials can reduce premiums. Insurance rates aren't set for each individual, but information about each individual is used to determine how much that person will pay. The amount an individual pays in income tax depends on a very specific personal attribute—how much money the person makes. People who make more money are not only charged more, but they are also charged at higher rates. Most people consider the progressive nature of our tax structure to be socially desirable.

In other situations, we may not be happy that we're getting a customer-centric price, but we know it's unavoidable. Whenever a negotiation is involved, each party attempts to determine the other party's willingness-to-pay. Good negotiators do so with great skill, but even those who aren't trained in negotiation can tell when a potential buyer is excited about the old wooden rocking chair at the garage sale. The casual banter of a good car sales agent—"What do you do for a living? Where do you live?"—is anything but casual. Through experience or training, good sales agents not only know how to close a sale, but when their commission depends on it, they also know how to generate the most possible revenue from each individual customer.

Even when we're upset about customer-centric pricing, there may be nothing we can do about it. Denise Katzman discovered that catalogue companies send different prices to different customers when she received a Victoria's Secret catalogue and found that her discounts were smaller than those of a male co-worker. Incensed, in 1996, she filed a class-action lawsuit. With no laws directly addressing Ms. Katzman's concerns, she contended that the differential pricing activity constituted mail fraud. The government's response was clear. U.S. District Court Judge Robert Sweet dismissed the case, ruling that "offering different discounts to different catalogue customers does not constitute mail fraud under any reading of the law," and that the "patently meritless complaint" caused "unwarranted adverse publicity" for Victoria's Secret. Katzman's attorney was sanctioned for filing a frivolous lawsuit and ordered to pay $5,000 toward Victoria's Secret's legal expenses.

While the majority of pricing science has focused on the product and not the customer, the scientific problems aren't insurmountable. With time and effort, customer-centric pricing can be refined in much the same way as product pricing. The biggest challenges, however, aren't related to the science.

On September 27, 2000, an article on the front page of the *Washington Post* read "On the Web, Price Tags Blur; What You Pay Could Depend on Who You Are." Amazon, a superstar of Web retailing, had been caught charging different prices to its customers. At question were DVDs. A customer who purchased a copy of Julie Taymor's *Titus* for $24.99 returned a week later to find that the price had risen to $26.24. Wondering whether the price change was specifically related to his buying habits, he cleaned his computer of cookies left by Amazon. Returning to the Amazon Web site, now as an unidentified visitor, he found a quoted price of $22.74. Don Harter, an assistant professor at the University of Michigan, had a similar experience. Accessing the Amazon Web site from two different computers, he found identical DVDs with prices that varied by more than $10.

Amazon denied using customer-centric pricing, declaring that the price variations were random. "It was done to determine consumer responses to different discount levels," said Bill Curry, a spokesperson for Amazon. Incorrectly referring to customer-centric pricing as dynamic pricing, he added, "This was pure and simple a price test. This was not dynamic pricing. We don't do that and have no plans ever to do that."

Still, visitors to the Web site DVDTalk.com weren't convinced. "They must figure that with repeat Amazon customers they have 'won' them over and they can charge them slightly higher prices since they are loyal and 'don't mind and/or don't notice' that they are being charged three to five percent more for some items," wrote one user. "This is a very strange business model," wrote another, "to charge customers more when they buy more or come back to the site more. I have no problem with coupons for first-time customers as marketing enhancements, but I thought the idea was to attract customers first and then work hard to keep them. This is definitely not going to earn customer loyalty."

Customer-centric pricing has long been held forth as one of the great promises of scientific retailing, and Internet retailing in particular. With computers to capture detailed transaction information, it seems a logical evolutionary step. As the incident with Amazon demonstrated, however, the Internet contributes to making customer-centric pricing *more difficult*. Amazon representative Curry understood this all too well. "[Customer-centric] pricing is stupid," he said, "because people will find out."

* * *

Acknowledging the applause of the crowd of over 300, Steve Pinchuk stepped to the podium at PROS's twelfth annual conference. Actually, he took the stage at the pricing conference, which was being held simultaneously with the revenue management conference in the next room. At one time only a single conference was held, but it quickly became clear that a dichotomy existed among the attendees. Everyone who attended was interested in the same issue—operational pricing. However, not everyone had been at it for the same length of time. Those with 10 or 20 years of experience in the travel industry were looking for different things than those who were just getting started on scientific pricing, not to mention that the travel industry, and airlines in particular, carry on discussions that revolve around reservations systems, global distribution systems, and inventory control. As a result, two separate conferences were formed. But people still mix in the halls, at meals, and at social events. Like all conferences, much of the learning takes place peer-to-peer, and it's always interesting when individuals from two entirely different spheres meet and realize just how much they have in common. Some attendees move back and forth between the two conferences, comfortable in both worlds.

Pinchuk is one of those people. Now corporate vice president of revenue management for Harrah's Casinos and Hotels, he spent most of his career involved with revenue management in the cruise industry, where he was introduced to the discipline by a former TWA executive. When Pinchuk joined Harrah's Entertainment in 2002, the company was already focused on revenue management thanks to executive leadership. If anything demonstrates just how far the gaming industry has come from its early days in the desert, it's Harrah's CEO, Gary Loveman. Holding a PhD in economics from the Massachusetts Institute of Technology, where he was an Alfred Sloan Doctoral Dissertation Fellow, Loveman started his career as an Associate Professor at the Harvard Business School. Intent on making things happen, he joined Harrah's in 1998 as the company's chief operating officer, taking over as CEO in January of 2003, when he continued to push forward with ambitious expansion plans. Loveman's biggest move came in mid-2005 with the acquisition of Caesar's Entertainment, making Harrah's the largest gaming company in the world with over $9 billion in annual revenues and more than 100 thousand employees. Caesar's Palace, Bally's, the Flamingo, and a half dozen other well-known names now belong to the Harrah's empire.

Harrah's success isn't linked just to its growth, but also to its focus on the customer. While many gaming companies spend exorbitantly on facilities, Harrah's invests in its customers. Making a huge investment in computer technology, the company developed a centralized repository of customer data, replacing the disparate systems associated with individual properties. Traveling to Las Vegas or Atlantic City, a customer who lives and gambles in Joliet, Indiana, now has a strong incentive to stay with Harrah's, earning and redeeming credits with the company's rewards program.

Cross-property loyalty, however, is only a small part of what customer data has to offer. As Pinchuk began his presentation, it almost seemed as if he were talking to the wrong audience; that his presentation should have been next door with the traditional revenue management practitioners. Casinos sell rooms, and Pinchuk was talking about room categories and inventory management. But it quickly became clear where he departed from traditionalists.

Casinos don't care about room revenues alone. Rooms make up an important portion of the overall revenue stream, but gaming, and increasingly entertainment, generate revenues that can far exceed the amount of money a customer spends on lodging. Recognizing this, Harrah's doesn't use room revenue to evaluate room price and availability, Rather, it uses estimates of the total revenue a customer is expected to generate. Customers that eat at expensive restaurants, attend performances, or lose consistently at the gaming tables can expect to find better room availability and prices. It's the age-old casino practice of comp'ing valued customers, but taken to a scientific level.

The efforts of Harrah's differ from the pricing experiments of Amazon in that Harrah's rewards customers for what they spend. Customers are not only comfortable with Harrah's pricing practices, but they also eagerly share information about their spending habits in order to be rewarded. There is no incentive to hide their identity. And the more they spend, the more they are rewarded.

Customer loyalty programs aren't new. Airlines pioneered the territory with frequent flyer programs, creating a new currency in the process. Where Harrah's has gone a step further is in scientifically linking customer-centric pricing into the loyalty equation. For companies that use loyalty programs, the question for science is to find the right equation.

Revolutionary as the efforts of Harrah's are, not every company uses loyalty programs. Science can still be applied, but to avoid problems similar to those encountered by Amazon, the focus must shift away from pricing to driving incremental sales using a collection of well-known techniques:

- *Up sell.* A customer divulges that he's looking for a computer to play video games, and an emphasis is placed on how each potential upgrade improves the game playing experience.
- *Cross sell.* Purchasing a book on juvenile crime in France during World War II, a customer is offered a list of similar titles and a discount on shipping when ordering more than one book.
- *Bundling.* Searching for a flight to San Francisco, a customer is offered the option of purchasing an all-in-one package including flight, hotel, and car.
- *Targeted marketing.* Acquisition of a fine Merlot is communicated to wine club members who have indicated a special interest in Merlots.
- *Discounts.* Frequent summer guests to a five-star hotel who have never visited during off-peak periods, George and Martha are offered a 25 percent price reduction and a free bottle of champagne if they visit on George's fiftieth birthday in October.

In each case, an attempt is made to transform information about the customer into incremental sales. Price is important, but it's secondary to the offer itself.

Customers aren't alienated with concerns about receiving a different price than their neighbors. And the better a company's science, the happier customers are to receive solicitations. Forgotten anniversaries may become a thing of the past thanks to unrestrained capitalism.

Whether it involves a country music performance, a massage at the spa, or uninterrupted time at the slot machines, Harrah's is positioned to offer its customers whatever they want most. Its goal is not just to create customer satisfaction, but also to drive higher revenues by providing something the customer is willing to pay for. Science is once again placing the customer at the center of the universe. We can only imagine what Copernicus would think.

* * *

"Price Management and Profit Optimization are the Best-Kept Secrets in Enterprise Software," reads the title of a July 2005 research report by the Boston-based Yankee Group. Adopting yet another title for pricing and revenue management—price management and profit optimization, or PMPO—the report opens with an equally unambiguous position statement:

> Now is the time for enterprises to invest in PMPO solutions. No other software delivers the same ROI: 10% to 20% profit improvement. In fact, the more broken and problematic your pricing processes are, the higher ROI you can expect. Leading companies that have already adopted PMPO solutions all have either met ROI expectations or—in many instances—exceeded them. Although pricing software offers the best demonstrated ROI compared to any other application software—beyond that of procurement, CRM, or ERP—it has yet to gain the attention or exposure on the scale of supply chain applications. It's been the best-kept secret of early adopters, many of which demand that their PMPO solution providers not sell the software to their competitors. None want to attest to the amazing ROI stories from their implementations for fear of losing a unique competitive advantage.

Research analysts like the Yankee Group play an interesting and important role in the modern economy. Fundamentally, their job is to spot important market trends then facilitate communication between companies that provide a good or service and companies that could benefit from using that good or service. It's a tough job, requiring extensive travel and the nurturing of a vast network of personal relationships—the bigger the better. On the positive side, research analysts can be as popular as publishers at a writers convention or federal grants program directors at an academic gathering.

Part of a research analyst's function is to categorize software in a way that defines a clearly articulated message to buyers and sellers in the market. It's a message that must empathize with the problems companies face while providing direction on where companies need to be going. The Yankee Group promotes the following three categories. Other analysts have similar categorizations.

Pricing-Analytics

Pricing-analytics software provides insights into pricing trends by presenting data quickly and in an easy to interpret format. It is related to the more general category of business intelligence software. Graphs and tables are used to show if the profit margin on a particular item is eroding, or to display information such as competitor or supplier prices. Pricing analytics software doesn't typically have a sophisticated science component, but rather provides information that isn't easily accessible in many organizations. Recognizing that your most demanding customer, after discounts and rebates, is generating negative margin is important to know, and fits comfortably into the business processes of most organizations.

Pricing Execution

Pricing-execution software focuses on making certain that the right prices are in place and enforced. Maintaining a master price list can be a difficult task, especially when companies sell thousands or tens of thousands of products. Coupled with constantly changing market factors such as costs of production, competitor prices, and a company's strategic pricing goals, basic price management can be a challenge. Once the prices are in place, it's important that the sales force is using them and is held accountable when they don't. Like pricing-analytics software, pricing-execution software doesn't typically employ a sophisticated science component.

Pricing Optimization

As the name suggests, pricing-optimization software is home to pricing science—the act of determining the prices that will be associated with items in inventory. Employing a broader definition than scientists use, pricing optimization incorporates forecasting and statistics in addition to optimization.

Yankee, AMR, Forrester, Gartner, and other research analysts are bullish about pricing and revenue management, and rightfully so. Sophisticated science isn't the entirety of their message. They know that science may or may not be the place to start. But science is an important part of the analyst's message as companies evaluate their long-term pricing strategies.

Over the last decade, I've seen and experienced many pricing successes firsthand. Joining a company that helped pioneer operational airline pricing at a time when it was taking its experience to new industries, I consider myself extremely fortunate to have participated in the process. I've seen ideas that were refined decades ago move seamlessly into new industries, and I've seen new ideas developed to address problems that airlines never encountered or have yet to consider. While there were inevitable missteps, long hours, and a lot of learning, there's no question that on the whole, it's been an exciting undertaking to understand the many pricing problems people face and to help bring a new way of thinking to the pricing process.

Like many who work on scientific pricing, however, I remain frustrated by the pace at which pricing and revenue management solutions are being adopted. In March of 2006, Forrester analyst Liz Herbert gave a presentation on the state of the market for pricing software and services. Referring to Forrester's experience in 2005, one slide read, "More inquiries from CFOs; More inquiries about optimization, not management; More inquiries from ERP vendors looking to round up their offering." The message of "more" wasn't difficult to miss, and the story continued. "Growing adoption rates; Bigger success stories; Higher profile pricing jobs; Deeper expansion in traditional verticals; New expansion in new verticals."

Another slide, however, carried a more sobering message. Depicting a bell-shaped curve on a time axis, it showed the technology adoption life cycle, with innovators and early adopters on the left, followed by the mainstream market in the middle, and laggards occupying the position under the right tail of the curve. Between the early adopters and the mainstream market, a portion of the curve had been removed, replaced by an arrow and the word "chasm."

"The greatest peril in the development of a high-tech market," read the slide, "lies in making the transition from an early market dominated by a few visionaries to a mainstream market dominated by pragmatists." Were pricing solutions in the chasm? Herbert left no doubt. "Yes!" read the slide, and she explained why. "Price optimization is still perceived as visionary, sophisticated; Still viewed as long, expensive implementation; Even with undeniable ROIs, there is very little penetration in the market."

Every new software technology faces impediments to its adoption. Buyers must be convinced that the technology saves or makes money, business processes need to be changed, and the software must be configured and installed so that it works properly. But scientific pricing software faces especially burdensome challenges. Based on mathematical data analysis, it impacts parts of an organization that have long relied on instinct and intuition. If the science is too complicated and users don't understand the results, the system is at risk of being rejected. Scientific pricing provides better decisions overall, but due to inherent uncertainty found in the world, it doesn't make perfect decisions every time. And pricing software must constantly concern itself with the practical limitations imposed by society's sense of justice.

It's understandable, then, that scientific pricing software is taking longer to gain broad acceptance than other enterprise software solutions. But with such a high return on investment, companies must eventually realize that scientific pricing is a necessary step to take. CFOs don't write a check without knowing where the money's coming from and how it will impact the company's financial status. Manufacturer's don't compete without understanding the available technology and how it can be used to keep production costs as low as possible. How is it that so many companies still price by staring at the ceiling until a number "feels right"?

* * *

The cover of the January 23, 2006, issue of *Business Week* looked strangely like an advertisement for a bad horror film. The mannequin-white face set against the blue background was covered with strange markings: deltas, thetas, omegas, cosines, logarithms, and integral signs. From the lips of the computer-generated figure came the words "WHY MATH WILL ROCK YOUR WORLD," and to the side, "More math geeks are calling the shots in business. Is your industry next?" An instant hit with the community of math geeks, e-mail messages flew, and copies of the picture and article began showing up on cubical walls. "Top mathematicians are becoming a new global elite . . . every bit as powerful as the armies of Harvard MBAs who shook up the corner suites a generation ago." Maybe, maybe not. But the thought was flattering to a group that rarely received such attention, even if they were being compared to Harvard MBAs.

As I discovered working with the airline industry, there are niches where sophisticated pricing science thrives. Airlines and the entire travel industry proved that scientific pricing could establish itself and, over time, flourish. Looking back, however, it's clear just how much effort it took to raise the industry from where it was to where it is. And even now, there remain important operational pricing issues that airlines haven't addressed.

As new industries begin to explore scientific pricing, their experiences will be much like those of the airlines 30 years ago. Yet this shouldn't deter them from moving forward. Boiled down to its most fundamental element, scientific pricing is about replacing guesswork with rational thought. Companies that implement even the most basic pricing initiatives will already be far ahead of their competitors. More importantly, they will be making a commitment to thinking rationally about their pricing. It doesn't take a PhD to start the process of scientific pricing. A lot can be done with addition, subtraction, multiplication, and division. If airlines are any indication of the future, however, once this commitment is made, companies won't be able to help but grow evermore sophisticated. "[The] mathematical modeling of humanity," concludes the *Business Week* article, "promises to be one of the great undertakings of the 21st century." As a math geek and business practitioner, I certainly like to hope so.

Notes

- For more on the Victoria's Secret case, see Weiss and Mehrotra (2001).
- Forrester analyst Liz Herbert's slides were presented at the 2006 PROS Pricing Excellence Summit, Houston, TX.

APPENDIX

Calculating Fare-Class Fares for Each Flight Leg in a Network

The problem of assigning a fare to each fare class on each flight leg is complicated when airlines operate a flight network of many interconnected flights. The process begins by solving a mathematical model similar to that used in the CFO example in Chapter 4. The model seeks to maximize revenue generated by the network while respecting constraints imposed by forecast passenger demand and the capacity of all the flight legs. The shadow prices (also discussed in Chapter 4) of the flight-leg capacity constraints in this model are used as the opportunity costs for each flight leg.

Pseudofares are then calculated using these opportunity costs. Figure A.1 depicts a simple network with four flight legs having one remaining seat each, and four passengers with their desired flight itineraries and associated Q-class prices. The opportunity costs generated from the mathematical model are depicted next to each flight leg. For each itinerary that makes use of the Chicago–Fargo flight leg, the pseudofare for that itinerary is calculated by taking the total itinerary fare and subtracting the opportunity costs of all flight legs *other* than the Chicago–Fargo flight leg. Thus, for example, the San Francisco–Chicago–Fargo pseudofare for the Chicago–Fargo flight leg is the total fare of $600 less the $1,000 opportunity cost of the San Francisco–Chicago flight leg. The three itineraries that use the San Francisco–Chicago flight leg each have a different pseudofare for this leg. Negative pseudofares are sometimes used, but are often replaced with a pseudofare of $0. Observe that the same itinerary will have a different pseudofare for each of the different flight legs that comprise it.

Once the pseudofares have been calculated for a flight leg and fare class, they are averaged to arrive at a representative fare. In our example, we calculate the average fare as follows:

$$1/3 \times \$200 + 1/3 \times \$400 + 1/3 \times (-\$400) = \$66.67$$

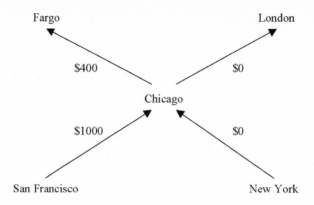

Itinerary	Passenger Demand	Q Fare	Chicago – Fargo Pseudofare
Chicago – Fargo	1	$200	$200
New York – Chicago – Fargo	1	$400	$400 ($400 - $0)
San Francisco – Chicago – Fargo	1	$600	–$400 ($600 - $1000)
San Francisco – Chicago – London	1	$1000	—

Figure A.1 Chicago–Fargo pseudofares for each of the Q-class itineraries using this flight leg.

The value $66.67 is used as the Q-class fare for the Chicago–Fargo flight leg when applying Littlewood's rule. In general, the weight applied to each pseudofare is proportionate to the expected demand for the associated flight itinerary.

Pseudofares are only one method of breaking down itinerary fares into flight leg fares. Other methods have been proposed and are used, but pseudofares are one of the most common and easiest to describe.

Reference List

Ascension Health. About ascension health. http://www.ascensionhealth.org/about/statistics.asp (accessed December 27, 2006).

Algers, S., and M. Besser. 2001. Modeling choice of flight and booking class—a study using stated preference and revealed preference data. *International Journal of Services Technology and Management* 2:28–45.

Alstrup, J., S. Boas, O. B. G. Madsen, R. Vidal, and V. Victor. 1986. Booking policy for flights with two types of passengers. *European Journal of Operational Research* 27:274–88.

Anderson, C. K. 2002. Revenue management: Issues in pricing and allocation. PhD diss., Richard Ivey School of Business, University of Western Ontario.

Andersson, S. E. 1998. Passenger choice analysis for seat capacity control: A pilot study in Scandinavian Airlines. *International Transactions in Operational Research* 5:471–86.

Aristotle. 1908. *Nicomachean ethics.* Trans. W. D. Ross. Oxford: Clarendon Press. See also the Web site of the Institute for Learning Technologies, 1995, http://www.ilt.columbia.edu/publications/artistotle.html.

———.*Rhetoric.* Trans. W. R. Roberts. Massachusetts Institute of Technology. http://classics.mit.edu/Aristotle/rhetoric.1.i.html, http://classics.mit.edu/Aristotle/rhetoric.2.ii.html, and http://classics.mit.edu/Aristotle/rhetoric.3.iii.html (accessed December 27, 2006).

Armstrong, J. S., and F. Collopy. 1992. Error measures for generalizing about forecasting methods: Empirical comparisons. *International Journal of Forecasting* 8:99–111.

Arrow, K. J. 1963. *Social choice and individual values.* 2nd ed. New York: Wiley.

Augustine. *De civitate dei. Patrologia Latina.*

———. *Ennaratio in Psalmum. Patrologia Latina.*

Baeck, L. 1999. The legal and scholastic roots of Leonardus Lessius's economic thought. *Storia del Pensiero Economico* 37, http://www.dse.unifi.it/spe/indici/numero37/baeck.htm.(accessed December 26, 2006).

Baker, S., and B. Leak. 2006. Math will rock your world. *BusinessWeek*, January 23.

Baker, T. K., and D. A. Collier. 1999. A comparative revenue analysis of hotel yield management heuristics. *Decision Sciences* 30:239–63.

Baldwin, J. W. 1959. Medieval theories of the just price: Romanists, canonists, and theologians in the twelfth and thirteenth centuries. *Transactions of the American Philosophical Society* 49 (4): 1–92.

Belobaba, P. P. 1987. Air travel demand and airline seat inventory management. PhD diss., Flight Transportation Laboratory, Massachusetts Institute of Technology.

Belobaba, P. P., and J. L. Wilson. 1997. Impacts of yield management in competitive airline markets. *Journal of Air Transport Management* 3 (1): 3–9.

Ben-Akiva, M., and S. Lerman. *Discrete choice analysis: Theory and application to travel demand.* Cambridge, MA: Massachusetts Institute of Technology Press, 1985.

Biehn, N. A cruise ship is not a floating hotel. *Journal of Revenue and Pricing Management* 5 (3): 135–42.

Bodily, S., and L. R. Weatherford. 1995. Perishable-asset revenue management: Generic and multiple-price yield management with diversion. *Omega* 23 (2): 173–85.

Bollapragada, S., H. Cheng, M. Phillips, M. Garbiras, M. Scholes, T. Gibbs, and M. Humphreville. 2002. NBC's optimization systems increase revenues and productivity. *Interfaces* 32 (1): 47–60.

Boodman, D. 1953. The reliability of airborne radar equipment. *Operations Research* 1 (2): 39–45.

Born, C., M. Carbajal, P. Smith, M. Wallace, K. Abbott, S. Adyanthaya, E. A. Boyd, et al. 2004. Contract optimization at Texas Children's Hospital. *Interfaces* 34 (1): 51–58.

Boyd, E. A. 1998. Airline alliance revenue management. *OR/MS Today* 25 (5): 28–31.

Boyd, E. A., and I. C. Bilegan. 2003. Revenue management and e-commerce. *Management Science* 49 (10): 1363–86.

Boyd, E. A., and R. Kallesen. 2004. The science of revenue management when passengers purchase the lowest available fare. *Journal of Revenue and Pricing Management* 3 (2): 171–77.

Boyd, E. A., E. Kambour, D. Koushik, and J. Tama. 2001. The impact of buy down on sell up, unconstraining, and spiral down. First Annual Meeting of the INFORMS Section on Revenue Management and Pricing. Columbia University, New York.

Bratu, S. 1998. Network value concepts in airline revenue management. Master's thesis, Massachusetts Institute of Technology.

Brumelle, S., and J .I. McGill. 1993. Airline seat allocation with multiple nested fare classes. *Operations Research* 41 (1): 127–37.

Brumelle, S., and D. Walczak. 1997. Dynamic airline revenue management with multiple semi-Markov demand. *Operations Research* 51 (1): 137–48.

Chvatal, V. 1983. *Linear Programming.* New York: Freeman.

Cook, T. 1998. SABRE soars. *OR/MS Today* 25 (3): 26–31.

Cooper, W. L. 2002. Asymptotic behavior of an allocation policy for revenue management. *Operations Research* 50 (4): 720–27.

Cooper, W. L., T. Homem-de-Mello, and A. J. Kleywegt. 2006. Models of the spiral-down effect in revenue management. *Operations Research* 54 (5): 968–87.

Copeland, D. G., R. O. Mason, and J. L. McKenney. 1995. Sabre: The development of information-based competence and execution of information-based competition. *IEEE Annals of the History of Computing* 17 (3): 30–57.

Copeland, D. G., and J. L. McKenney. 1988. Airline reservations systems: Lessons from history. *MIS Quarterly* 12 (3): 353–70.

Cross, R. G. 1997. *Revenue management: Hard-core tactics for market domination.* New York: Broadway Books.

Curry, R. E. 1990. Optimal airline seat allocation with fare classes nested by origin and destinations. *Transportation Science* 24:193–204.

Dantzig, G. B. 1998. *Linear programming and extensions*. New edition. Princeton University Press.

Davenport, T. H. 2006. Competing on analytics. *Harvard Business Review* (January): 98–107

Davis, P. 1994. Airline ties profitability to yield management. *SIAM News* 27 (5): 12, 18–19.

De Lollis, B., and B. Hansen. 2005. Airlines give fliers fewer chances to do the bump. *USA Today*, December 19.

Duke University. Admissions and enrollment statistics, the graduate school of Duke University, PhD only. http://www.gradschool.duke.edu/about_us/statistics/admitece .htm (accessed December 26, 2006).

Economagic. All grades conventional retail gasoline prices. http://www.economagic .com/em-cgi/data.exe/doewkly/day-mg_tco_us (accessed December 27, 2006).

Eklund, J. 1994. The reservisor automated airline reservation system: Combining communications and computing. *IEEE Annals of the History of Computing* 16 (1): 62–69.

Elizondo, R., E. A. Boyd, and M Beauregard. 1997. Evaluating school facility capacity and attendance boundaries using a large-scale assignment algorithm. *Economics of Education Review* 16 (2): 155–61.

Elmaghraby, W., and P. Keskinocak. 2002. Dynamic pricing in the presence of inventory considerations: Research overview, current practices, and future directions. *Management Science* 49 (10): 1287–309.

Faaland, B. H., and F. S. Hillier. 1979. Interior path methods for heuristic integer programming procedures. *Operations Research* 27 (6): 1069–87.

Feldman, J. M. 1990. Fares: To raise or not to raise. *Air Transport World* 27 (6): 58–59.

Fetter, R. B. 1991. Diagnosis related groups: Understanding hospital performance. *Interfaces* 21 (1): 6–26.

Foulkes, J. D., W. Prager, and W. H. Warner. 1954. On bus schedules. *Management Science* 1 (1): 41–48.

Gallego, G., and G. J. van Ryzin. 1994. Optimal dynamic pricing of inventories with stochastic demand over finite horizons. *Management Science* 40 (8): 999–1020.

Gass, S. I. 1990. *An illustrated guide to linear programming*. Reprint edition. New York: Dover.

Gilmer, R. W. 2005. Houston after the hurricanes. Report by the Federal Reserve Bank of Dallas. October. http://www.dallasfed.org/research/houston/2005/hb0507.html (accessed December 27, 2006).

Glover, F., R. Glover, J. Lorenzo, and C. McMillan. 1982. The passenger mix problem in the scheduled airlines. *Interfaces* 12:73–79.

Grossman, G. A. 2003a. Getting down to business: The changing landscape of AACSB accreditation presents new opportunities and challenges for OR/MS at business schools. *OR/MS Today* 30 (4).

———. 2003b. Up to standard: How should management science faculty respond to the new accreditation guidelines? *OR/MS Today* 30 (4).

Hillier, F. S., and G. J. Lieberman. 2004. *Introduction to operations research*. 8th ed. New York: McGraw-Hill.

Horner, P. R. 2002. History in the making: INFORMS celebrates 50 years of problems, solutions, anecdotes and achievement. *OR/MS Today* 29 (5): 30–39.

Huang, K. 2005. Price management and profit optimization are the best-kept secrets in enterprise software. Research report by the Yankee Group, Boston, MA.

Kantorovich, L. V. 1975a. Autobiography submitted to the 1975 Nobel Prize Committee. http://nobelprize.org/nobel_prizes/economics/laureates/1975/kantorovich-autobio
.html (accessed December 24, 2006).

————. 1975b. Mathematics in economics: Achievements, difficulties, and perspectives.
Nobel Prize acceptance lecture, December 11. http://nobelprize.org/nobel_prizes/
economics/laureates/1975/kantorovich-lecture.html (accessed December 24, 2006).

Karaesmen, I.Z. 2001. Three essays on revenue management. PhD diss., Columbia University.

Karlaftis, M. G., K. G. Zografos, J. D. Papastavrou, and J. M. Charnes. 1996. Method-
ological framework for air-travel demand forecasting. *Journal of Transportation Engineering* 122 (2): 96–104.

Kimes, S. E. 1999. Group forecasting accuracy for hotels. *Journal of the Operational
Research Society* 50 (11): 1104–10.

————. 2005. Restaurant revenue management: Could it work? *Journal of Revenue and
Pricing Management* 4 (1): 95–97.

Kimes, S. E., and L. W. Schruben. 2002. Golf course revenue management: A study of tee
time intervals. *Journal of Revenue and Pricing Management* 1 (2): 111–20.

Ladany, S. P., and D. N. Bedi. 1977. Dynamic rules for flights with an intermediate stop.
Omega 5:721–30.

Lautenbacher, C. J., and S. J. Stidham. 1999. The underlying Markov decision process in
the single-leg airline yield management problem. *Transportation Science* 33:136–46.

Lee, A. O. 1990. Airline reservations forecasting: Probabilistic and statistical models of
the booking process. PhD diss., Massachusetts Institute of Technology.

Lee, T. C., and M. Hersh. 1993. A model for dynamic airline seat inventory control with
multiple seat bookings. *Transportation Science* 27:252–65.

Levinson, M. 2001. Jackpot! Harrah's big payoff came from using IT to manage customer
information. *CIO Magazine*, February 1.

Littlewood, K. 1972. Forecasting and control of passenger bookings. *AGIFORS Annual
Symposium Proceedings* 12:95–117. See also *Journal of Revenue and Pricing Management*
4 (2005): 111–23.

Markowitz, H. M. 1952. Portfolio selection. *Journal of Finance* 7 (1): 77–91.

McCartney, S. 2000. Bag of high-tech tricks helps to keep airlines financially afloat. *Wall
Street Journal*, January 20.

McGill, J. I. 1993. Censored regression analysis of multiclass demand data subject to joint
capacity constraints. *Annals of Operations Research* 60:209–40.

McGill, J. I., and G. J. van Ryzin. 1999. Revenue management: Research overview and
prospects. *Transportation Science* 33:233–56.

Migne, P. L., ed. *Patrologia Latina*. 1844 to 1855, 1882 to 1865. http://pld.chadwyck
.co.uk/.(accessed December 26, 2006).

Orkin, E. 1998. Wishful thinking and rocket science: The essential matter of calculating
unconstrained demand for revenue management. *Cornell Hotel and Restaurant and
Administration Quarterly* 39:15–19.

Patsuris, P. 2001. Delta's no Priceline savior. *Forbes*, February 9.

Petzinger, T, Jr. 1995. *Hard landing: The epic contest for power and profits that plunged the
airlines into chaos.* New York: Three Rivers Press.

Popescu, I. 1999. Applications of optimization in probability, finance and revenue management. PhD diss., Massachusetts Institute of Technology.

Rao, B. V., and B. C. Smith. 2006. Decision support in online travel retailing. *Journal of Revenue and Pricing Management* 5 (1): 72–80.

Richter, H. 1982. The differential revenue method to determine optimal seat allotments by fare type. *AGIFORS Annual Symposium Proceedings* 22:339–62.

Robinson, L. W. 1995. Optimal and approximate control policies for airline booking with sequential nonmonotonic fare classes. *Operations Research* 43 (2): 252–63.

Rosen, C. 2001. Southwest Airlines files suit against Orbitz. *Information Week*, May 4. http://www.informationweek.com/story/IWK20010504S0009 (accessed December 27, 2006).

Rothkopf, M. H., and S. S. Oren. 1979. A closure approximation for the nonstationary M/M/s queue. *Management Science* 25 (6): 522–34.

Rothstein, M. 1971. An airline overbooking model. *Transportation Science* 5:180–92.

Saint Patrick's Catholic Church, Washington, D. C. Louise de Marillac, widow. http://www.saintpatrickdc.org/ss/0315.htm#loui (accessed December 27, 2006).

Sandoval, G. 2001. Ticket squabble besets net travel industry. *CNET News.com*, March 30. http://news.com.com/2100-1017-255050.html?legacy=cnet (accessed December 27, 2006).

Schachtman, N. 2000. Trading partners collaborate to increase sales. *Information Week*, October 9. http://www.informationweek.com/807/cpfr.htm (accessed December 27, 2006).

Secomandi, N., T. Atan, K. Abbott, and E. A. Boyd. 2002. From revenue management concepts to software systems. *Interfaces* 32 (2): 1–11.

Seton Healthcare Network. The charity care and community benefit report for 2004. http://www.seton.net/about_seton/charity_care/charity_report_2004_final.pdf (accessed Decmber 27, 2006).

Shapiro, C., and H. R. Varian. 1998. Versioning: The smart way to sell information. *Harvard Business Review* (November–December): 106–14.

Shaykevich, A. 1994. Airline yield management: Dynamic programming approach. Master's thesis, Department of Operations Research, University of North Carolina.

Smith, A. 2003. *The wealth of nations*. New York: Bantam Classics. Originally published in Edinburgh, 1776.

Smith, B. C., D. P. Gunther, B. V. Rao, and R. M. Ratliff. 2001. E-commerce and operations research in airline planning, marketing, and distribution. *Interfaces* 31:37–55.

Smith, B. C., J. F. Leimkuhler, and R. M. Darrow. 1992. Yield management at American Airlines. *Interfaces* 22 (1): 8–31.

Smith, B. C., and C. W. Penn. 1988. Analysis of alternative origin-destination control strategies. *AGIFORS Annual Symposium Proceedings* 28.

Streitfeld, D. 2000. On the web price tags blur: What you pay could depend on who you are. *Washington Post*, September 27.

Subramanian, J., C. J. Lautenbacher, S. J. Stidham. 1999. Yield management with overbooking, cancellations and no shows. *Transportation Science* 33:147–67.

Sun, X. 1992. Airline yield management: A dynamic seat allocation model. Master's thesis, Faculty of Commerce, University of British Columbia.

Svrecek, T. 1991. Modeling airline group passenger demand for revenue optimization. Master's thesis, MIT Flight Transportation Laboratory, Massachusetts Institute of Technology.

Talluri, K. T., and G. J. van Ryzin. 1998. An analysis of bid-price controls for network revenue management. *Management Science* 44 (11): 1577–93.

————. 1999. A randomized linear programming method for computing network bid prices. *Transportation Science* 33:207–16.

————. 2004a. Revenue management under a general discrete choice model of consumer behavior. *Management Science* 50 (1): 15–30.

————. 2004b. *The theory and practice of revenue management.* Dordrecht, The Netherlands: Kluwer Academic Press.

Tertullian. *De idolatria. Patrologia Latina.*

Thorp, E. O. 1961. A favorable strategy for twenty-one. *Proceedings of the National Academy of Sciences* 47 (1): 110–12.

————. 1966/1962. *Beat the dealer: A winning strategy for the game of twenty-one.* 2nd ed. New York: Vintage.

Tobin, J. 1958. Estimation of relationships for limited dependent variables. *Econometrica* 26:24–36.

Train, K. E. 2003. *Discrete choice methods with simulation.* Cambridge: Cambridge University Press.

Travel Industry Association. Fun research statistics. http://www.tia.org/researchpubs/stats.html (accessed December 26, 2006).

U.S. Department of Health and Human Services, Centers for Medicare and Medicaid Services. U.S. national health care expenditure highlights. http://www.cms.hhs.gov/NationalHealthExpendData/downloads/highlights.pdf (accessed December 27, 2006).

Vazsonyi, A. 2002. Reminiscences & reflections: My first taste of OR, "I had a helluva big assignment." *OR/MS Today* 29 (5): 36.

Vinod, B. 2006. Advances in inventory control. *Journal of Revenue & Pricing Management.* 4:367–81.

Weatherford, L. R. 1991. Perishable asset revenue management in general business situations. PhD diss., Darden Graduate School of Business Administration, University of Virginia.

Wei, Y. J. 1997. Airline O-D control using network displacement concepts. Master's thesis, Massachusetts Institute of Technology.

Weik, M. H. 1964. A fourth survey of domestic electronic digital computing systems. Ballistic Research Laboratories Report 1227, Aberdeen Proving Ground, Maryland.

Weiss, R. M., and A. K. Mehrotra. 2001. Online dynamic pricing: Efficiency, equity, and the future of e-commerce. *Virginia Journal of Law and Technology* 6 (2).http://www.vjolt.net (accessed December 27, 2006).

Whitaker, A. 2004. CRS deregulation: Just the mention of it draws cheers and jeers from travel industry. *Tampa Bay Business Journal,* January 23.

Williamson, E. L. 1992. Airline network seat inventory control: Methodologies and revenue impacts. PhD diss., Massachusetts Institute of Technology.

Wilson, J. L. 1995. The value of revenue management innovation in competitive airline industry. Master's thesis, Massachusetts Institute of Technology.

Wolfe. B. D. 1964. *Three who made a revolution.* 4th revised ed. New York: Dell.

Wollmer, R. D. 1992. An airline seat management model for a single leg route when lower fare classes book first. *Operations Research* 40 (1): 26–37.

Wolverton, T. 2001. Orbitz hit with trademark lawsuit. *CNET News.com,* May 4. http://news.com.com/2100-1017-257076.html?legacy=cnet (accessed December 27, 2006).

Zeni, R. H. 2001. Improved forecast accuracy in airline revenue management by unconstraining demand estimates from censored data. PhD diss., Rutgers University.

Zhang, D., and W. L. Cooper. 2005. Revenue management for parallel flights with customer-choice behavior. *Operations Research* 53 (3): 415–31.

Zhao, W. 1999. Dynamic and static yield management models. PhD diss., Wharton School, Operations and Information Management Department, University of Pennsylvania.

Zhao, W., and Y. S. Zheng. 2001. A dynamic model for airline seat allocation with passenger diversion and no-shows. *Transportation Science* 35 (1): 80–98.

Zickus, J. S. 1998. Forecasting for airline network revenue management: revenue and competitive impacts. Master's thesis, Massachusetts Institute of Technology.

Index

academic community, 10, 19–20, 122, 126, 130–32, 141
aeronomics, 20
Air New Guinea, 16
air traffic, 9, 121
airline industry: costs in, 15, 31, 112, 125; delays in, 121; fare classes in, 11–13; fare-class inventory levels in, 11, 55, 61; forecasting in, 46; frequent flyer programs of, 170; history of, 23–35, 41n; as indicators of the future of scientific pricing, 4, 174; inventory control in, 44, 61, 64; Internet and, 14; nested inventory levels in, 46; performance measures in, 121; pilot training in, 144; reforecasting in, 63; reputation of, 115; reservations systems in, 26–32; revenue management in, 15, 21, 57, 61, 65n, 83n, 88, 117–21; sales departments in, 32; setting fares in, 34–35; unconstraining demand observations in, 79, 83n
airline pricing, 2–5, 7–8, 35, 91, 115, 123, 126, 166–67, 177; changes in, 63, 113; comparison to an auction, 87, 113; comparison to game of marbles, 45; comparison to gas pricing, 35; comparison to hotel pricing, 90; consumer discomfort with, 3, 9, 104, 112–13; group bookings, 62–63; marginal revenue and, 47
Airline Tariff Publishing Company, 28
airlines. See *individual airlines*
Albert the Great, 110
Alexander the Great, 104
Amazon, 168, 170
American Airlines, 15, 20, 23–30, 34–36, 119–21, 125, 132
American Airlines Decision Technologies (AADT), 125
American Automobile Association (AAA), 90
American Society of Travel Agents (ASTA), 28–29
AMR, 20, 119, 125, 173
Apollo, 27–28, 30. *See also* PARS
Aquinas, Thomas, 110
Aristotle, 104–6, 108, 110, 113
Arrow, Kenneth, 132
athletic events, 93
Augustine, 107–8, 116

Baldanza, Ben, 22, 35
Baldwin, John, 105–7

Belobaba, Peter, 19
bid price, 47
blackjack, 70–75, 77, 117
BlueLinx, 138–46, 149
bookings, 11–12, 27–29, 39, 85, 90;
    electronic, 15; fictitious, 33; global
    distribution systems and, 86;
    group, 62–63, 100; history of,
    23–26; origin of the term, 23;
    patterns in, 3, 101; reservations
    system and, 31, 36. *See also*
    overbooking
Botone, Bernard of Parma, 110
Bowman, Dave, 123
Braniff International Airways, 30
Brunger, Bill, 13–15, 22, 34–38
bundling, 170
Burr, Donald, 36, 41
business passengers, 9, 11–12, 44
business schools, 129
BusinessFirst seats, 38
buy-down, 79, 81, 83

Caesar's Entertainment, 169
capitated contracts, 158
carve outs, 159
case rate contracts, 158–59
casino industry, 71–72, 169–70
CFO example, 49–56, 59, 61, 177
checkerboarding, 91
Civil Aeronautics Board, 11, 13, 30
Coca-Cola Company, 113–14
co-hosting, 30
Coleridge, Samuel Taylor, 153
commission, 2, 28, 33, 100, 149, 167
common sense, 18, 68, 118
communication, 88, 105–7, 119–26,
    132–33
competitor price information, 80–81,
    172
computer literacy, 140
computer reservation systems, 2
Consolidated Reservations Control
    (CRC), 32

consumer choice, 81, 83
Continental Airlines, 8, 10, 13, 19,
    32–33, 35–40, 94, 98
contracts, types of, 157–59
Control Data Corporation, 28
Cook, Thomas, 119–21, 125–26, 132
Copernicus, Nicolaus, 165, 171
Copernican Revolution, 165
Corbett, Neil, 99–101
cost-to-serve, 139–40, 144
Crandall, Bob, 26, 28, 29, 36, 120–22,
    125–26
Cripe, Dave, 160
cross sell, 170
Curry, Bill, 168–69
customer-centric pricing, 165–70
customer loyalty programs, 170
customer relationship management
    (CRM) software, 165–66, 171
CyberKnife, 157

Dantzig, George, 54
data, availability of, 4
data cleansing, 149
Daughters of Charity of Saint Vincent
    de Paul, 153–54
Davis, Lee, 41
Delta, 20, 26, 29
DELTAMATIC, 26
demand curves, 148–53, 162
deregulation, 2, 11, 14, 32, 34, 36, 43,
    125
DFI, 20
diagnosis related groups (DRGs), 159
DINAMO (Dynamic Inventory
    Allocation and Maintenance
    Optimizer), 35–36
discount-off-charges contracts, 158–59
discounts, 11, 13, 93, 158, 167–68,
    171, 172
displacement cost, 47
Douglas, Michael, 43
duality theory, 55
Dudziak, William, 138–42, 145

Eastern Airlines, 8, 23, 27, 29
Edelman Prize, 129–31, 159
Einstein, Albert, 147
elevator pitch, 122
Emirates Airlines, 9
engineering schools, 120, 129
Evler, John, 154–57, 160, 162
Expedia, 86, 88
"explanation," understandings of the
    word, 124
Exxon Mobil, 111

fares, setting, 34–35
Fetter, Robert, 159
flight legs, number of, 91–92
Florida Express, 7–8
Fargo (movie), 3
forecasting, 3–4, 8, 20, 39, 63, 68–71,
    79, 81, 172
Forrester, 173
Frontier Airlines, 7
fuel costs, 15

Galilei, Galileo, 165
Gallego, Guillermo, 130
GammaKnife, 157
Gartner, 173
General Motors, 130
Georgia-Pacific, 138, 140, 142–43
global distribution system, 31, 86, 169
golf industry, 92
group bookings, 62–63, 100

halo effect, 29
Harrah's Entertainment, 169–71
Harter, Don, 168
health-care industry, 153–63
health insurance, 154–56
Herbert, Liz, 173
Hertz, 86
hotel industry, 85; mathematical
    similarities to revenue
    management, 90–91
Hold 'em, Texas, 67–68, 75, 83

hub-and-spoke systems, 15, 57
Hughes Aircraft, 127
hurdle rates, 47, 90–91
hurricanes, 111–12

IBM, 26, 37
Indigent Care Collaboration, 154
Institute for Operations Research and
    the Management Sciences
    (INFORMS), 10, 117–19, 122,
    128–31
Interfaces (journal), 128
International Air Transport Association
    (IATA), 9
Internet: data availability and, 4; effect
    on pricing of, 14–15; reservations
    systems and, 31; travel industry
    and, 86–87
inventory control, 7–13, 17, 32–35,
    39–44, 64, 79, 169
inventory levels, 20, 32, 38–39, 61, 63;
    nested, 46–47, 56, 58
Ivester, M. Douglas, 114

Joint Industry Computerized Reserva-
    tions System, 28
just profit, 109, 112
justice, 105–115
Justinian Code, 108

Kantorovich, Leonid, 51–54
Katrina, hurricane, 111
Katz, Jeff, 120
Katzman, Denise, 167–68
Kimes, Sheryl, 97
Kinloch, Leon, 41
Koopmans, Tjalling, 54
Kubrick, Stanley, 123

laesio enormis, 108–110
Le, Tuan, 67, 75–77, 82
ledgers, 23–24
legacy carriers, 14–15
leisure passengers, 11–12

length-of-stay, 91
Lenin, Vladimir, 51, 53, 56
Lenin Prize, 53
Lennon, Vernon, 138–39, 141–42
Little, John, 148
Little's Law, 147
Littlewood's rule, 48, 56, 57–59,
    63–64, 77–78, 178
load factors, 18, 34, 43–44
Lorenzo, Frank, 8, 19, 37
Lough, Greg, 41
Louise de Marillac, 153
Loveman, Gary, 169
low-cost carriers, 14–15
Ludwig, Al, 24–26, 34, 36

Macadam, Stephen, 143–45
Magnetronic Reservisors, 2, 25
management science, 118, 127–28. See
    also operations research
Management Science (journal), 54,
    130–31
Marathon Oil, 111
marble game, 45
marginal costs, 89, 144
marginal revenue, 47–48, 55, 112
marginal value, 54–55
market price, 110–13
Markowitz, Harry, 132
married segment logic, 92
Marx, Karl, 51
mathematical models, 49–51, 59, 61,
    81, 92–93, 113, 133, 177
Maxfield, Paul, 67, 76, 82
MBAs, 2, 40, 87, 129, 174
McElrea, Charles, 142–43
Medicaid, 154, 155
Medicare, 154, 155, 159
Melchior, Mariette, 19
mercantile activity, 106–9
Midway Airlines, 7
Morse, Philip, 126–28

National Airlines, 7
National Science Foundation, 3, 129,
    131
nested inventory levels, 46–48, 56, 58
network inventory valuation, 62
Newton, Isaac, 165
Nicomachean Ethics, 110
Nimoy, Leonard, 87
Nobel Prize, 51, 53–54, 127, 132
nonprofit organizations, 148, 153–55
normalization, 149
Northwest Airlines, 3

objective function, 50
Official Airline Guide, 28
oil industry, 111–12
operational pricing, 2, 3, 8, 13, 16, 21,
    34–36, 41, 80, 169, 174
operations research, 117–20, 122–23,
    125–32
Operations Research (journal), 127–28,
    131
Operations Research Society of
    America (ORSA), 126–28
opportunity costs, 47, 58–64, 90–91,
    100, 177
"optimal," definition of, 48
optimization, 54–55, 63, 118, 135,
    161, 172–74
Orbitz (travel Web site), 14, 62, 86,
    87, 120
overbooking, 10, 18, 36, 38, 89

Pan Am, 7, 26
PANAMAC, 26
Papua New Guinea, 15–16
PARS (Programmed Airline
    Reservations System), 26–27
passenger information, 23, 26, 27
per diem contracts, 158
Philippine Airlines, 9
Pinchuk, Steve, 169–70
Plato, 104–6, 108
poker, 67–68, 75–76, 82–83

price management and profit
    optimization (PMPO), 171
price response function, 148–49
Priceline, 87
pricing analysts, 34–35
pricing-analytics software, 172
pricing-execution software, 172
pricing-optimization software, 172
problem solving, 118, 127, 131,
    135–37
profit management, 100
profit, 10, 13, 21, 106–116, 148, 156,
    158, 171–72
profitability, 4, 5, 43–44, 52–60
Propsys (PROS), 3, 8, 16–22, 32, 87,
    95, 123–24, 137, 141–43, 169,
    175
pseudofares, 58–59, 177–78

Quillinan, John, 102

randomness, 70–72, 75, 77, 118, 138,
    144, 149
refundability, 13, 166
regional carriers, 30
relationship building, 17
rental car industry, 30, 85, 88–91
reservations systems, 26–31, 36–41,
    45–46, 61, 63, 86, 89, 92, 125.
    See also global distribution systems
restaurant industry, 93, 95–98, 136–37
revenue equation, 148, 162; abuses of,
    149–53
revenue management, 57, 61–65, 80,
    83, 123, 131–32, 149, 163,
    169–71; adoption of solutions for,
    173; American Airlines and,
    34–40, 121, 132; business
    characteristics of, 88–89;
    customer-centric pricing
    compared to, 166–67; definition
    of, 8–10; Littlewood's rule and,
    48; marble game analogy for, 45;
    mathematics and, 90–98; pricing

problems and, 89–90; software,
    19–21; transition from inventory
    control, 34
Rickenbacker, Eddie, 23
Rita, hurricane, 111
Roman Catholic Church, 106–111,
    165
Roman law, 108–9
Royal Dutch Shell, 111

SABRE Decision Technologies, 125
SABRE reservations system, 20, 26, 30,
    125, 133
sales agents, 2, 4–5, 100, 138–45, 167
sales simulator, 144
Salter, Robert, 8, 16–21, 32–34
science, selling, 19–20
scientific pricing, 4–5, 137–38, 143;
    applications of, 130, 161–62, 169;
    changing field of, 37; frog in the
    pot analogy, 18–19; future of, 4,
    119, 133, 174; goal of, 82; secrets
    of, 35, 81; software for, 173–74;
    uncertainty and, 64; visibility of,
    126
screen bias, 29–31
secretary problem, 44, 56, 68
Seton, Elizabeth Ann, 153
Seton Healthcare Network, 153–57,
    160, 162
shadow prices, 52–56, 177
Shafer, Mark, 94–101
Shatner, William, 87
Sheen, Charlie, 43
Simon of Bisignano, 110
Singapore Airlines, 9
Smith, Adam, 104
Smith, Cyrus Rowland, 23, 26
Socrates, 104
Sousa, Manny, 39, 41
Southwest Airlines, 13, 88
Soviet Union, 51–54
Stalin, Joseph, 52

stochastic processes, 118
Stone, Oliver, 43
stop loss, 159
strategy, 71–78
Sweet, Robert, 168
Swissair, 120
system dynamics, 144

targeted marketing, 170
Taymor, Julie, 168
Teamsters, 32
technology, 38, 40, 115, 125–26, 140,
    145, 169; adoption cycle of,
    173–74
Teleregister Corporation, 25
Teleregister Telefile, 25–26
Tertullian, 107
Texas Hold 'em. See Hold 'em, Texas;
    poker
Texas International Airlines, 8, 30
The Institute for Management Sciences
    (TIMS), 127–28
theory of one, 17
theory-versus-practice debate, 126–32
Thorp, Edward, 71–72
Tiffany system, 24
time-to-delivery, 12–13
travel agents, 14, 27–31, 38, 40,
    85–86, 92

Travelocity, 86

unconstraining, 79, 81, 83
unions, labor, 10, 32–33, 43
up sell, 170
U.S. Airways, 14, 35, 131

van Ryzin, Garrett, 130–31
Vazsonyi, Andrew, 127
venture capital, 87
venture capitalists, 122
Victoria's Secret, 167–68
Vincent de Paul, 153
von Neumann, John, 54

Wall Street (film), 43–44
Walt Disney World, 93–101
wealth, 106–7
Winemiller, Bert, 123–24
Woestemeyer, Ronald, 8, 16–21, 32
World Poker Tour Championship
    (2005), 67

Yankee Group, 171–72, 173
yield management, 8, 36. See also
    revenue management
Yuen, Benson, 7–9, 16–18, 20, 21